POSITIVE PSYCHOLOGY FOR MUSIC PROFESSIONALS

Positive Psychology for Music Professionals is a guidebook to the building blocks of positive psychology and character strengths, and the ways in which they can be used by music professionals throughout the industry to empower, celebrate, and leverage individuality.

Written in a highly accessible and entertaining tone – and based on the science of character pioneered by the VIA Institute – this book is designed to introduce the language, themes, and concepts of a strength-based approach to working in the music industry. Targeted exercises, self-reflections, interviews, and profession-specific case studies encourage readers to harness the power of their strengths to shift to an open mindset, create more positive working relationships, and improve institutions within their field.

Positive Psychology for Music Professionals is essential reading for music professionals of all kinds, including aspiring and established musicians, students, music producers, educators, and managers in all sectors of the industry.

Raina Murnak is an assistant professor and director of Contemporary Voice and Performance Artistry and Program Director of the MA in popular music pedagogy at the University of Miami's Frost School of Music. She's a thought leader on stage presence, voice, movement, and strength-centered pedagogies.

Nancy Kirsner PhD, TEP, OTR has been in private practice, teaching, and consulting for 45 years. A certified positive psychologist, she writes about translating positive psychology into action methods. Nancy is co-author of "Positive Psychology and Psychodrama" in *Action Explorations: Using Psychodramatic Methods in Non-Therapeutic Settings* (2019).

"If music is the language of the Gods, then this book is about helping music professionals realize their full humanity and, by doing so, create heaven on earth. A compelling read for all of us who savor one of life's greatest gifts."

Tal Ben-Shahar, *Founder of Happiness Studies Academy & Potentialife*

"This book exposes the beautiful and shiniest gem in music education. It reveals and sheds light on that inner knowing that we have all felt for so long navigating through our personal musical journeys. After reading this book, you will have a deeper relationship with yourself that will be beautifully expressed through your artistry."

Cassandra Claude, *Author and Vocal Coach*

"This book is an absolute delight. Beautifully written, it is a feast of creativity [...] Although this book is, as the title professes, an introduction to the character strengths for music professionals, anyone from any profession will gain a great deal by reading it. Not only do we learn a tremendous amount about the character strengths, we also get an inside scoop into the world of music. I highly recommend this book to anyone living in a human body wishing to add more insight, joy, and inspiration to their lives!"

Megha Nancy Buttenheim, *Let Your Yoga Dance Founder and Positive Psychologist*

Positive Psychology for Music Professionals

Character Strengths

RAINA MURNAK AND
NANCY KIRSNER

Routledge
Taylor & Francis Group

LONDON AND NEW YORK

Designed cover image: © Raina Murnak

First published 2024
by Routledge
4 Park Square, Milton Park, Abingdon, Oxon OX14 4RN

and by Routledge
605 Third Avenue, New York, NY 10158

Routledge is an imprint of the Taylor & Francis Group, an informa business

British Library Cataloguing-in-Publication Data
A catalogue record for this book is available from the British Library

Library of Congress Cataloging-in-Publication Data
Names: Murnak, Raina, 1979, author. | Kirsner, Nancy, author.
Title: Positive psychology for music professionals : character strengths /
Raina Murnak and Nancy Kirsner.
Description: Abingdon, Oxon ; New York : Routledge, 2023. |
Includes bibliographical references and index. |
Identifiers: LCCN 2023022706 (print) | LCCN 2023022707 (ebook) |
ISBN 9781032212746 (paperback) | ISBN 9781032212760 (hardback) |
ISBN 9781003267607 (ebook)
Subjects: LCSH: Musicians--Psychology. | Positive psychology.
Classification: LCC ML3838 .M9473 2023 (print) | LCC ML3838 (ebook) |
DDC 781.1/1--dc23/eng/20230515
LC record available at https://lccn.loc.gov/2023022706
LC ebook record available at https://lccn.loc.gov/2023022707

ISBN: 978-1-032-21276-0 (hbk)
ISBN: 978-1-032-21274-6 (pbk)
ISBN: 978-1-003-26760-7 (ebk)

DOI: 10.4324/9781003267607

Typeset in Giovanni
by MPS Limited, Dehradun

Access the Instructor(s) and Student Resources: www.positivepsychologyformusicprofessionals.com

CONTENTS

FIGURES

TABLES

ACKNOWLEDGMENTS

We are so grateful to so many people and experiences who have brought us to the point of writing this book.

Raina's Acknowledgments

I would like to thank my Frost School of Music family and the University of Miami for awarding me the Arts and Humanities Fellowship to support this work. Thank you to our editors Hannah Rowe and Emily Tagg at Routledge for believing in this project and being so helpful along the way.

Thanks to the many students who have been a part of this project from the early forays into applying positive psychology teachings and practices, to the pilot study, and to the hundreds of students who have volunteered to take the VIA survey and empower their strengths to work toward their goals.

Thank you to all the music professionals who have taken the mindful time to participate in this book and share so many special parts of your stories. It was so amazing to hear about your journeys through the lens of your strengths. I learned so much more this way about people I have known and worked with for years! Seeing the ways in which they thrive by using their powers inspires me to cultivate more of my own underused strengths.

Thank you also to my family, friends, and loved ones who have heard the phrase, "sorry I'm writing the book" far too often. I owe you. Thank you, Dawn Murnak-Karageorgos, for not only being my sister and biggest fan, but being a role model wise beyond her years, who is a beacon of positivity and light. Thank you, Maria and Richard Murnak, for supporting me in every step of my often left-of-center journey and for encouraging me to keep reaching for all of the stars. Thank you, Brian Wilkins, for providing me with wisdom and insight into so many parts of the music industry.

A very special thank you to Pharrell Williams, who not only embodies these teachings in every aspect of his being, but also gives so much back to youth, students, and the music community through his constant educational generosity and his striving to make the world a more beautiful and empathetic place.

Thank you finally to Nancy, my partner in writing, who not only nurtured this baby with me but has helped me grow in so many ways through the years. This is our true labor of love.

Nancy's Acknowledgments

What a blessing to have so many people to thank for making this book come to fruition. This book wouldn't have happened without my co-author,

Dr. Raina, inviting me to write it with her. We had been on a five-year adventure together learning and experimenting with character strengths in the world of musical professionals and how to best elevate their educational experience. This book has been a pure expression of her creativity, love, patience, and joyful living. Yes, we actually had a lot of fun doing this together.

Without my positive psychology community, beginning with Dr. Tal Ben-Shahar, Megan McDonough, Megha Nancy Buttenheim, Maria Sirois, and Phoebe Atkinson, I would not even know the basic concepts, and more specifically that the VIA character strengths even exist. Phoebe Atkinson is often my intellectual muse and heart person, showing me new paths of knowledge along with her sprinklings of love and gratitude. Megan has been a role model of creativity and leadership in action – she embodies the best of both. Megha taught me about anchoring my knowledge in the wisdom of the body and how to move with my strengths. Maria brought spirituality and inspirational teaching into my life and changed the way I present forever. With Dr. Tal at the intellectual helm, I had the best blend of learning – academic, the classics, literature, the newest research, and the important people in the field.

To my sisters and brothers from the Certificate in Positive Psychology Program, thank you for sharing yourselves as we learned, laughed, and told stories together over the years. Lena Lisitskaya and Nicole Stottlemyer have inspired me with their vitality, joy, and passion for life. All of you have been a remarkable support and resource.

Thank you to Dr. Ryan Niemiec and the VIA Institute for their steadfast commitment to teaching and researching the VIA strengths and their applications and for giving us permission to use the VIA Classification in our work. (© Copyright 2004–2023 VIA Institute on Character. All rights reserved. Used with permission. Viacharacter.org.)

The ongoing courses and creative ways to use strengths are always there to support and encourage the use of character strengths.

My clients and groups, over a 40-year period, have been some of my best teachers – thank you. Your authenticity, vulnerability, and desire to grow have continually inspired and touched me. You have provided a rich interpersonal field for us to explore.

I want to thank my psychodrama trainers, Dr. Dale Richard Buchanon and Dr. Nina Garcia, for many years of mentoring around experiential learning and translating the tools and skills of psychodrama into practical teaching tools. Once they put me into action – I never taught the same way again. Lulu Carter is a creative genius of psychodrama and all the expressive arts. Lulu has more love than you can imagine and she shares it generously and lifts me to another celestial level; thank you, Lulu.

My Temple Beth Or Community and especially my friends, Marcy Prince, Phylis Winnick, Sheila Silverberg, Gail Ironson, Sara Horowitz, and Martha Rothaus, were there to support me and help me not take myself or things too seriously!

And lastly, there is my family. My daughter, Shira, and her husband, Joe, and my granddaughters, Annie and Izzi, have waited patiently when Grammy was busy working on the book. My son, Scott, and his wife, Amy, both professional writers, and grandchild, Ollie, have cheered me on and are always supportive. My assistant, Caridad Montane, is an angel and she has my back every step of the way. This is my village – my tribe – and I thank each of you for being in my life.

GREEN ROOM

Forewords and Thoughts from Our Colleagues

Raina Murnak and Nancy Kirsner

Green Room

1. *noun*: a room (as in a theater or studio) where performers can relax before or after appearances; a place where performers are not onstage or on camera. Friends and family of the musicians are also invited into the green room.

Pharrell Williams
Producer • Songwriter • Performer • Designer • Entrepreneur • Genius • Empath

I have learned through my experiences as a student of Dr. Raina Murnak firsthand that she is an empath, she is a connector, and she is a liberator.

As an empath, she senses a person's presence below the skin, straight to the core, right in the spirit. She sees who you are from your outer shell to whom you want to be in your spirit. And then she makes a connection. She's a master of creating connections. She meets you at the intersection of your learning journey where you're at and sees how far you can go at that juncture. She's right there, right in the intersection with you. And it might have stop lights and people speeding – it doesn't matter. If you are standing there, she's going to meet you and assure you that you're in a safe space, where comfort and confidence are waiting on you to just fully embrace.

And then she's a liberator. When you come to the other side of seemingly tough conversations or conversations that take time to understand, she's ushering you to this place where you realize, "Okay, it wasn't really that tough. And it really didn't take that much time." It took what it needed to get you where you needed to be. To be liberated from the baggage.

I experienced this firsthand with myself. And I witnessed this firsthand with a third-party artist we both had encountered. Raina and I both saw the same thing in this individual – we saw that they were locked. They were locked by an assortment of traumas- some self-imposed, some environmentally imposed, and others developing because of happenstance situations. We could both clearly see that this person was hitting a wall. And it was interesting to me to see because of the way that I unlock those scenarios for folks and give them clarity when in music studio sessions, she had the same vision. She also had very similar tactics, processes, and methods for unlocking, connecting, and showing a person what their freedom just might look like, feel like, and sound like.

It's a combination of three things. It's not just what you're seeing. It's not just what you're hearing. But it's also what they're feeling. She's a master of empathy, connection, rapport, and liberation because she uses no less than these three senses to concoct the right elixir that is designed for each individual; to get inside.

Empathy is the skeleton key that opens those doors. If you want to understand how to open the lock, you must first become the lock and understand it. And once you do, then you have the key. And once you make many, many, many keys, over and over, then you become the locksmith of confidence, and comfort, electrifying oneself, and showing others how to access their deepest heat, their strongest electricity.

This has been my experience with Dr. Murnak and I'm sure I'm not alone. There are a lot of other people who have been unlocked, who have been felt and understood, and who have been on the receiving end of that connection. And who are happy to walk around with their degrees that describe the paths they took in their educational journey. But the symbolism of that degree is that they have been liberated.

Kate Reid
Jazz Vocalist • Educator • Director of Jazz Vocal Performance

Music professionals and creatives of all types constantly work toward finding then honing their artistic voice, their personal sound or style, and their purpose in what can often feel like a very dense landscape that is the music industry. Individuals in the arts look to refine the special aspect or aspects of their talent or skill that separates them from others focusing on the same goal. This can be an intense or arduous journey to gain the skills or develop qualities in our performance that will either change our career trajectory or advance us to the next level of opportunity. That intensity can make or break the artist resulting in a climb to the top or a spiral down and away from our creative and musical goals altogether.

The mental game for sustaining a career in music or the arts requires a tenacious spirit and a steadfast belief that you have something special to offer. The truth is we all have qualities and attributes within our own spirit and character that make us unique and can help drive our success, creativity, and artistic growth. However, our traditional music training doesn't always provide for the development of positivity in our own thinking or recognition nor the application of our own personal character strengths.

As a professional musician for more than twenty-five years, I have experienced my fair share of self-doubt and low self-worth throughout my own musical journey which has led from time to time to periods of low creativity and artistic growth. I'm certain many of my peers and colleagues experience these momentary valleys as well and perhaps we are led to believe the successful musical journey requires this grind. As an educator for over 20 years, I have witnessed in many students the lack of self-assuredness and pride in what they bring to the musical table. At times, even those receiving awards in their field or for solo opportunities in the academic setting are lacking confidence and experience anxiety about their artistic output.

In their book, *Positive Psychology for Music Professionals: Character Strengths*, Dr. Raina Murnak and Dr. Nancy Kirsner show us the importance of a strength-centered approach to music and working within the arts and provide us with an understanding of the value of our own uniqueness. The book gives an overview of the science of character strengths and basic concepts in positive psychology. Raina and Nancy explore the 24 character strengths and share the research from their pilot study of music professionals of all types. They offer exercises and methods to apply our strengths in a variety of settings including performance, classroom and ensemble education, the music industry, and therapeutic use.

I recently had the opportunity to take the VIA survey and meet with the authors about my own character strengths and mindset as it relates to creativity. Although we only touched the surface of this process, I immediately gained insight into how I perceive my own artistic place in this musical world, which of my character strengths have brought me to the place I am currently in my creativity and how I might utilize more of my untapped strengths as I move forward with my next creative project scheduled to begin later this year.

Raina and Nancy have created an accessible and entertaining approach to the examination of positive psychology and character strengths for music professionals, artists, students, and educators alike. The research and methods in this resource affirm we can adjust our mindset and create pathways to build upon and implement the strengths unique to each of us. I am adopting *Positive Psychology for Music Professionals: Character Strengths* as required reading for my voice studio and I highly recommend this amazing resource for all artists and music professionals no matter your experience or where you are on your musical or career journey!

Megan Mcdonough

Founder • Wholebeing Institute

When you begin using positive psychology tools – the evidenced-based practices for human flourishing – life opens up in unexpected ways.

I've seen it again and again over the past decade leading Wholebeing Institute. When people take 9 months to learn, apply, and practice the tools of positive psychology, change happens. The culmination of that experience is the final project – a presentation about what is most compelling and interesting for each student.

The range is as varied and unique as each person. Because the program focuses on what is most important for personal growth, not a test or a paper to prove proficiency, there's freedom to explore. People write poems or stories. Create a game for a daughter with special needs. Build a family ritual of counting blessings at dinner time. Produce an album of original songs about the best of humanity. Develop a course about resilience for stressed-out college or high school students. Start a new travel business that focuses on savoring and positive emotions. Reimagine a spiritual community. Reorient life to heal after trauma, illness, or death.

Nancy Kirsner choose to sing with Raina Murnak.

Years later, the unexpected result is the book you now hold in your hands.

To my knowledge, this book is the first to specifically and clearly apply the science of positive psychology character strengths for the music professional. Reading it connects you to the humanity behind the music. Making music is more than technique. Music comes out of this perfectly imperfect thing called being human. There's a universality of fear – stage fright, vulnerability, insecurity. And there's a universality in counteracting it – bravery, perseverance, honesty.

Learning character strengths help you tune your human instrument, so what comes out as music is an expression of the whole you.

Nancy's final project engaged the character strength of bravery, stepping out of her comfort zone to give voice – literally – to that which elevated her. You'll hear more about Nancy and Raina's story later in the book. For now, I want to hit home the importance of this book for your own music. The science of character strengths, often called the backbone of positive psychology because of the foundational and essential role they play in thriving, means you'll learn to sing or play (and live) from the core of who you are – your most authentic way of being.

The strength of your voice in singing (and in life) comes from knowing the strength within – the character strengths that enliven you and your music. That's the promise and the power within these pages.

Singing teachers will show you how to use your voice. Raina and Nancy teach you how to embrace your humanity, using your own character strengths for more ease, less stress, and greater satisfaction.

May you play your brightest song always, living in the incredible gift of this crazy and surprising human life.

SOUNDCHECK

Preface Chapter

Raina Murnak and Nancy Kirsner

Soundcheck

1. *noun*: a test of sound equipment; an on-the-spot rehearsal by a band before a gig to enable the sound engineer to set up the mixer.

Testing ... Check 1,2,1,2 ... Turn Up Your Character Strengths!

A good sound check leaves performers confident and ready, sure of their equipment and levels. This crucial preparatory phase also allows the performers to get comfortable with the layout of the stage and feel of the room. A successful sound check leaves both the performer and engineer feeling confident and ready for showtime. In this sound check, we want you to begin to get comfortable with the empowering language and concepts of character strengths.

In this section you will be guided to take a survey and discover your unique and personalized character strength profile. This survey of 24 positive character strengths is expressed differently in everyone.

Character strengths are one of the main foundations of contemporary positive psychology, and an important research-based pillar in the science of whole being (or happiness). This branch of science deals with evidenced-based practices that offer accessible tools of enrichment for all areas of life. Just as positive psychology itself was a response to the negative bias of traditional psychology, focusing on your character strengths will shift your mindset from "what's wrong with you" to "what's strong in you." Our highest-level character strengths are also energizing and a lot of fun to explore!

As musicians and music professionals, we aim to be creative and authentic and approach our work with zest. And yet, our formal training tends to overly focus on our weaknesses rather than our strengths and sometimes the initial joy can get lost. This kind of negative lens can feel overwhelmingly critical and antithetical to building the inner confidence and personal truth creativity requires. Research has shown that working with a strength-based approach builds confidence, lessens anxiety, and brings out one's authentic voice.

This book vivifies the teachings of positive psychology and character strengths using direct applications for music professionals. We teach you to recognize ways your unique strengths are already expressing themselves in your work and offer activities to boost the strengths you wish to cultivate.

And most importantly, as with a great sound check, we hope at the end of your journey you can truly hear yourself.

Backstage

About the Authors

Backstage

1. *noun*: the area behind the proscenium in a theater, especially in the wings or dressing rooms. Out of view of the public; in private; behind the scenes.

Nancy Kirsner, PhD, TEP, OTR, CPP

Traditional Biography

Most of my professional life, I was introduced with a typical bio like this:

Nancy Kirsner, PhD, TEP, LMFT, OTR, began her career in helping others as an occupational therapist in a VA hospital during the Vietnam War. Over the last 40 years, she has worked as a Gestalt art therapist, licensed marriage and family therapist, university professor, and trainer and practitioner of psychodrama and group psychotherapy. With a PhD in counseling psychology, Nancy has worked as a forensic consultant with attorneys, and she facilitates bibliodrama (stories from the Bible) at temples and churches. Nancy is the co-author of the chapter *Positive Psychology and Psychodrama* in *Action Explorations: Using Psychodramatic Methods in Non-Therapeutic Setting* with Phoebe Atkinson.

So that is my typical bio. Now let me introduce myself to you through the world of character strengths! I took my first VIA in 2014 and since then I have taken it four more times (in 2016, 2017, 2020, and 2021) as I was curious about the reliability and stability of my strengths and how situational and emotional events impacted them. I am going to use the 2021 version that was taken in the second year of COVID.

Now meet me through my song of strengths and its top five notes – my signature strengths. Which do you prefer?

Character Strength Biography

Spirituality (Transcendence) I was always a spiritual person, and during this time spirituality was for the first time my number one strength. The multitude of life-changing experiences around COVID and the fact that I was facilitating two open groups a week (over 350 to date) online, constantly had me thinking and listening to others asking about the purpose and meaning of their lives which had so drastically been blown up. We were trying to find our place, where we belong in this uncertain changing world.

Creativity (Wisdom) Creativity was my second strength, and it has always been in my top six strengths. Creativity is so core to me, it is my everlasting light – I am always thinking of new ways to teach and do things. I visualize things as I am relaxing, and I am always full of ideas on how to do something different whether in my home, garden or as a writer. As a therapist who teaches, creativity is always by my side – I can easily think of another way to say things or walk through another door.

Gratitude (Transcendence) I am so excited to have gratitude as my third character strength. Gratitude for me is the heavy hitter of all my strengths and gives me more bang for my buck than anything else. Gratitude connects me to all mankind and if I'm feeling down, it takes me right to my heart place, seeing all beings as part of God unifying us all as family. Expressing and feeling grateful is like breathing to me and it calms my soul. It is my go-to strength.

Fairness (Justice) Fairness is currently my fourth character strength – it is always in my top and sometimes causes me distress in life. It is very important to me that all people are treated fairly and without bias. Of course, every day we all see examples of people being treated unfairly. Even as far back as when I was a little girl, I didn't understand why this upset me so much until I learned about my strengths. It was very helpful for me to learn this as it tempered my responses to unfairness.

Humility (Temperance) Wow! This was a welcomed change that I worked on for several years – humility is number five – my last signature strength. In earlier years, humility was in the lower third of my character strengths. Humility is a Temperance strength and most folks I have worked with ask me "what does that word mean?" I had a well-hidden idea that I was more special than others and my accomplishments were very important to me. Now, I don't need to tell everyone who I am, or what I have done. It is an easier, gentle way to live.

What do you think … which one gives you a better sense of who I am? One is about "doing" and accomplishments – more the "outside" me. The character strength introduction says more about "being" or the inner me.

Raina Murnak, DMA

Character Strength Biography

Creativity (Wisdom) I was raised on the value of doing things differently. Whether it was creating dramatic fashion moments, decorating my locker like a miniature living room, or asking my father to help me build a high reading perch in my room, I was keenly aware of my need to push past the norm to find the more whimsical version of what life could be. This creativity was a testament to my artistic parents, especially my father, Richard Murnak, who encouraged me to be out of the box. He came from advertising and coming up with creative concepts was almost a holy value in my household. As I grew as an artist and professor, I extended this value into all performances and classes. The novelty and fun I bring to teaching voice, performance artistry, and music theory proved to be successful in reaching multiple types of learners and those intimidated by the subject itself.

Appreciation of Beauty and Excellence (Transcendence) I have deep emotional responses to human excellence, thriving, and striving. The beauty of human achievement, the awe of nature, and the gift to be alive – these are at the absolute forefront of my thoughts every day. They are not in the background or something I have to remember to find – they are my raison d'etre! Making a space beautiful, interesting, and inspiring – be it my office, home, or even computer screen – is my mandatory starting place before I can begin a day of work. As the director of Contemporary Voice and Performance Artistry at the Frost School of Music and as a performer myself, great attention to staging, costuming, lighting, and audience experience factors at the forefront.

Zest (Courage) Zest is everything to me if you couldn't tell by my bio already! As an empath, I'm highly aware of the way energies impact us and I take full responsibility for the energy I bring to the room, a project, a class, a brunch- you name it! When I feel at my best, I bring my zest, tightly tied to my humor. The more mundane, the more I aim to zestfully make it fun. Taxes, you say? Tax-spa I say as I work poolside on receipts. Final voice jury, you say? Final performance for all your friends to celebrate you as an artist! There is always a way to make things more fun, more infused with love and joy.

Humor (Transcendence) Humor is something I never leave home without. I couldn't get through one day without reframing all that occurs in some sort of curbed enthusiasm. Seeing the funny version, or the characters that surround me brings such lightness and transcendence to any situation. I mean, there have been many a wardrobe malfunction, many an on- or offstage anxiety attack, and many an epic fail at work- I have to laugh at these, put them in perspective and move on. Reframing these otherwise traumatic moments this way takes away their negative, draining power. I'm always actively thinking about what is funny about a moment or what has funny potential.

Curiosity (Wisdom) And thankfully, my curiosity keeps bringing me to new places and new challenges and even this book! I also am writing the music for a Broadway-bound musical as I write this book. I had never done anything like this or that before, but my curiosity dared me.

When I first experienced learning about my own strengths, I was in an "artistic crisis." I was in the middle of a reinvention period professionally and I sort of felt I was at the end of my creative rope. Through working with Nancy, reminding myself, and reaffirming who I was at my core, I was not only able to get through my own blockages but flourish to help others with theirs. On an almost-weekly basis, I find myself helping students with similar artistic crossroads transitions. Having them tap into their personal strengths empowers them to embody their truth, dare to chase their dream, and value the currency of their uniqueness and originality. The moment I stopped trying to be a poor copy of "them" and started being a priceless one-off of myself, my life changed.

Traditional Biography

Dr. Raina Murnak is an assistant professor of contemporary voice and performance artistry for the Frost School of Music at the University of Miami where she has been innovating in higher education for the past 19 years. She is the program director of Frost Summer Institute of Contemporary Songwriting and the MA in popular music pedagogy for FrostOnline. She gives master classes as a thought leader on stage presence, popular voice, movement, and positive strength-centered pedagogies. Dr. Murnak was the co-director of the 2021 FMEA Popular Music Collective and the 2022 NAfME All-National Honor Ensemble for Modern Band. Recently, she co-starred off-Broadway with Emmy-nominated actress Janine Turner in the original musical they wrote, *Belva* (music Murnak/book and lyrics Turner). She holds both a BM and MA degree in vocal performance and composition from the SUNY at Stony Brook, a DMA in music theory/composition and vocal pedagogy from the University of Miami, and a certificate in e-learning instructional design and development at Oregon State University.

Which version gave you more of a sense of who I am?

Our Hopes for You, the Reader It is our sincerest hope to introduce you to the splendor and power of your own character strengths. We want you to leave this book feeling reaffirmed, reinvigorated in your musical and professional pursuits, and proud of the unique perspective and abilities you add to every situation by simply being yourself.

Stage Plot

How to Use This Book

Stage Plot

1. *noun*: a visual representation that illustrates a band or performer's live performance setup, band member placement on stage, what gear you use, and any other helpful information to the venue.

This book introduces a main pillar of positive psychology – the well-researched practice of knowing and engaging one's character strengths. It is meant to be user-friendly and colloquial to musicians while teaching important concepts in positive psychology and neuroscience as a foundation for understanding and applying one's character strengths. To get the most out of this book, please first take the VIA strengths survey at www.viacharacter.org to see your very own strengths profile. Everyone's strength profile is unique to them – this is the excitement of this exploration – there is only one YOU!

You will use this to look at yourself through a strengths-based lens and learn ways to leverage these strengths specifically in your chosen musical endeavor or profession. Everything included in this book from positive psychology – the concepts, neuroscience, and the sciences of whole being, and character strengths are research-based and useful to all aspects of your life, even if you haven't yet chosen a specific musical path. At the end of this chapter, you will find a chart where you can write down all 24 of your strengths and refer to them throughout the course of this book.

There are opportunities to experientially integrate the concepts and suggestions of each chapter in the final PRACTICE ROOM sections. These are dedicated to furthering growth, application, and understanding of your strengths as you go along your journey. Positive psychology and character strengths are part of a rich resource of practices – to really make things stick, the action part is essential.

In Chapter 1, OPENING ACT, you will get a foundational understanding of the development of character strengths followed by some research highlights. You will learn about key concepts including the aware-explore-apply model, strength spotting, strength blindness, and using them in balance. We will describe our pilot study and how it impacted our participant's experiences then and into the future. You will also get a tour of your amazing brain, as this book will connect some neuroscience basics to enrich and deepen your understanding of character strengths from a larger perspective.

Chapter 2, FIRST SET, provides an in-depth look at each individual strength, grouped by virtue category. Each of these strengths will include definitions, examples, the neuroscience and brain connections of each strength, workbook areas for self-reflection and writing, activities to boost your strengths, and a profile showcasing a musical professional who embodies the spirit of that strength. This is a fun section and a unique way to better understand yourself and your social or professional circle. Read about these strengths for yourself and think about the people you interface with (friends, family, bandmates, co-workers, students), as well as yourself, to really get the most out of this section.

Chapter 3, SECOND SET, brings in the larger virtue categories in which these strengths are grouped and relates them to career paths in music. In this chapter, you will hear from professionals leveraging their strengths to create successful, satisfying lives and livelihoods. You can use this as both an analytic and exploratory chapter. You will be able to delve into professions and learn which strengths they are fueled by. Paired with that, interviews with music professionals will exemplify how each person specifically uses their unique set of strengths in those professions and paths. You can also "matchmake" your own strength profile to the categorized professions. Designing a career path and life that taps into your top strengths guarantees you will be energized, engaged, and at your best.

Chapter 4, LAST SET, demonstrates strength-based activities and ideas for your purposes and populations to explore and apply strengths. Once you have learned about the power of using character strengths, these action interventions will help you integrate what you have learned and put them to work in your life. This chapter includes the ENCORE which will bring your character strength odyssey to a close and emphasize the power of always leaning on your strengths, especially during trying times. The AFTER PARTY and MERCH BOOTH will provide additional resources and products to help you with your strength-based practices.

We hope to empower and revitalize you with the strengths you already possess!

Table 0.1 Character Strength Worksheet

Grouping	#	Strength
TOP 5 SIGNATURE STRENGTHS	1	
	2	
	3	
	4	
	5	
HIGHEST 8 STRENGTHS (Include Top 5)	6	
	7	
	8	

(Continued)

Table 0.1 (Continued)

Grouping	#	Strength
MIDDLE 8 STRENGTHS	9	
	10	
	11	
	12	
	13	
	14	
	15	
	16	
LOWER 8 STRENGTHS	17	
	18	
	19	
	20	
	21	
	22	
	23	
	24	

Soundcheck Practice Room

Identify where and how your signature strengths are showing up in your life.

You are encouraged to do this activity socially with friends and family. Sometimes we don't see our highest strengths because we assume they are just common to everyone and therefore not particularly special. Our friends and family can help us view the strengths we are "blind" to because they are so natural and normalized to us.

_____
_____
_____
_____
_____
_____

CHAPTER 1

Opening Act

The Science of Character Strengths

Opening Act

1. *noun*: The first act at a concert before a main act. The opening act performance serves to "warm up" the audience, making them excited and enthusiastic about the headliner. It also helps the audience transition from their external realities into the "here-and-now" moment of the concert experience.

Why Character Strengths

Character strengths are some of the most important aspects of an individual's personality – especially the signature or top five strengths. They are the basic building blocks of goodness, and core parts of our best selves. They are the main tools of positive psychology used throughout this book. The power of character strengths is what inspired us to write this book. Learning your character strengths will give you a lens and language of positivity that will shift your mindset from wrong to strong. In other words, they will facilitate moving you into a growth/open mindset. This helps you become a better, more resilient, and more confident version of yourself.

Character strengths are expressed in degrees, shaped by context, and show up differently in each situation. In other words, they have structure, depth, and dimensionality. Character is individualized and idiosyncratic and everyone has a unique profile of character strengths. At the same time, these character strengths are interdependent – they work in combination with each other – like muscle groups – and each impacts the other. One can think of this in the same way multiple notes create a harmony or chord or multiple layers of instruments or voices create the sound color of a band, choir, or orchestra.

DOI: 10.4324/9781003267607-1

People have the fundamental need to express who they are, be seen, valued, and accepted, and to feel part of something greater than themselves. Part of what makes the character strengths so important and exciting is their ability to concretize and give language to these vital parts of us. Learning your character strengths is something that will forever change you, your mindset, and the way you think and talk about yourself and others. This will be a significant transition in your life, both personally and professionally.

The aim of this book is to teach you to recognize and apply these powerful energizers! Knowing and understanding your character strengths (and those of others in your life) will help you make more informed choices in your professional pursuits and your relationships. It's a game changer inclining you towards happiness, well-being, greater engagement, and more creativity.

Your Signature Strengths

Positive psychology research refers to one's top five strengths as signature strengths. These indicate the strengths you are currently using most in your life – what you value and how it shows up in your behavior in the world. They are essential to your identity, like your name or written signature. All the literature and research on character strengths use this top-five signature strength model. For our own purposes and to see a wider context, we have found it useful to look at the 24 character strengths grouped as the highest eight, middle eight, and "growing edge" lower eight. Naming high, middle, and low groups of our character strengths help us see them as more fluid and connected than ordinal rank numbers from 1 to 24.

Signature strengths have three key defining qualities, which we call the three "E's":

- They feel *essential* to who you are.
- They feel *effortless* and natural.
- They are *energizing*, creating balance and happiness.

Basic research findings around signature strengths show that when you use, work, and live in your high-level character strengths, you experience greater happiness and vitality in your life. They bring a higher quality to our performance, and we feel more engaged and invigorated. As you become aware of your signature strengths, you will feel authentic ownership of them, and others will easily recognize them in you (Peterson & Seligman, 2004).

While your five signature character strengths can be relatively stable over time, they can also shift based on the situational demands of your life or because of your intentional practices. As well, the fact that character strengths are plural, interact, and influence each other also produces changes.

The Value of Our Lesser Strengths

Our lowest eight character strengths (the growing edge) don't come as naturally in our lives as other higher ones. Remember though, lesser strengths are not weaknesses! It is common that as soon as someone receives their ranked order list from their VIA Survey, they look to the bottom of the list to see what is least ranked – what they are not good at. This is our natural human negativity bias in action. The work of the VIA Institute and research demonstrates that we can build these lesser strengths through intentional practice and mindfulness skills that leverage the power of our stronger character strengths to reach our goals.

The Science and Research of Character Strengths

In the early 2000s, Doctors Martin Seligman and Chris Peterson developed a common language of 24 VIA character strengths. VIA stands for "values in action." These strengths are bravery, honesty, zest, perseverance, prudence, self-regulation, forgiveness, humility, curiosity, creativity, love of learning, perspective, judgment, hope, humor, gratitude, appreciation of beauty and excellence, spirituality, love, kindness, social intelligence, fairness, leadership, and teamwork.

While people have many more than 24 strengths, we all commonly have these selected strengths in varying degrees. This is part of what makes us so unique. The character strengths that were finally chosen for the VIA survey, were most universal across all cultures on Earth and throughout the written history of time. The founders of the VIA project organized the strengths into six categories: courage, temperance, wisdom, transcendence, humanity, and justice.

The project took researchers three years to identify and choose which would be the 24 character strengths for the survey. The list was published in the book *Character Strengths and Virtues: A Handbook and Classification* (Peterson & Seligman, 2004). The VIA character strengths are regarded as the "backbone of the science of positive psychology" ("Character strengths and virtues: A handbook and classification," n.d.). The VIA Institute was established as a nonprofit organization in Cincinnati, Ohio, that supports VIA education, activities, and research. It was founded in 1999 by Dr. Neal Mayerson (PhD in clinical psychology) and Dr. Martin E. P. Seligman (PhD and former president of the American Psychological Association). Since its inception, Dr. Ryan Niemiec, Psy. D. has served as the educational director of the VIA Institute.

Character strengths are a prominent aspect of positive psychology. The VIA organized and operationalized a universal vocabulary for the language of strengths. Historically, psychiatry, psychology, and education have primarily focused on what's wrong with people. The lens of the character strengths changes our perspective from "what's wrong" to "what's strong." Character strengths highlight what is good and strong in people while creating an appreciation of our differences. Learning and applying character strengths leads to the ability to recognize other people's strengths as complementary to a whole.

There are many positive uses for this in the musical context, whereby so much of musical training focuses solely on "what's wrong," often ignoring all of what is already strong. You will see many examples of ways you can apply strength-focused approaches to music professions and activities.

The VIA Classification and VIA Inventory of Strengths (VIA-IS) are widely used around the world by researchers and practitioners – it has been taken by over 25 million people worldwide! The research on character strengths is robust and the number of studies has grown consistently since the initial publication of the VIA Classification system in 2004. Most of this research has been published in peer-reviewed scientific journals.

The VIA-IS has been translated into over 40 languages. There is a youth assessment for children/adolescents ages 10 through 17. Globally, the most reported character strengths are kindness, fairness, honesty, gratitude, and judgment. The least-reported character strengths are prudence, humility, and self-regulation ("VIA Inventory of Strengths (via-IS)," n.d.).

The VIA-IS has good reliability and validity. Character strengths are mostly stable but do change in response to life events, or deliberate lifestyle actions. Aristotle himself endorsed an approach to living whereby one's character strengths are expressed in a golden mean – or in balance (Niemiec, 2019).

Applied research about character strengths is new territory. While there is a wealth of character strength practices, strength-based models, applications, and interventions, the practice of character strengths has not been studied extensively. There is a strong movement of applying the research on character strengths across disciplines, as is the case with this very book, applying character strengths to the world of music professionals.

Research Highlights

Here are some highlights of the most well-established, salient, and recent research on the benefits of recognizing your character strengths. A high-level summary of the research shows several important links between character strengths and life satisfaction, engagement, wellbeing, meaning, better health behaviors, less stress, more positive emotions, and increased achievement and performance. These are all desired outcomes that have far-reaching benefits to one's whole being and musical life.

Early Research

The first decade of character strength research focused on establishing a universal language about what is best in people. This opened the door to important principles and laid the groundwork for the science of character (Niemiec, *Via character strengths: Research and practice (the first 10 years)*, 2013).

Enhanced Well-Being and Life Satisfaction

This is the most common variable studied in positive psychology and character strengths. Another word for this is *happiness*. Using your signature strengths contributes to greater well-being and less psychological distress in adults (Linley et al., 2010).

In another study, possessing these five character strengths showed a consistent relationship to life satisfaction: hope, zest, gratitude, curiosity, and love. The character strengths least related to life satisfaction were found to be humility, creativity, appreciation of beauty and excellence, judgment, and love of learning (Park et al., 2004). This shows the strengths of the "heart" (love, gratitude) are more strongly associated with well-being than the strengths of the "head" (love of learning, judgment) ("Developing character strengths," 2019).

Achievements in School and at Work

Perseverance is the leading character strength related to success at work and in school. Other character strengths that predicted GPA in college students were love of learning, humor, fairness, and kindness (Lounsbury et al., 2009).

There are also positive correlations between positive moods at school and the character strengths of perseverance, social intelligence, zest, and love of learning (Weber et al., 2016).

Another study exploring connections between the use of signature strengths, goal progress, psychological needs, and well-being found that those who used their signature strengths made more progress on their goals, met their basic psychological needs, and were higher in overall well-being (Linley et al., 2010).

Law students exhibited greater satisfaction and a decreased likelihood of depression and stress by using their signature strengths (Peterson & Peterson, 2008).

Improvement of Mental Health and Psychological Problems

There is considerable improvement of mental and psychological problems when you understand what your character strengths do and how they show up in your life.

Developing your signature strengths can play a role in reducing depression (Malouff, John, 2019).

Specifically, the character strengths of hope, zest, and leadership are related to fewer problems with anxiety and depression. Hope, kindness, social intelligence, self-regulation, and perspective buffer against the negative effects of stress and trauma (Park 2006, 2008a, 2009a).

Self-acceptance

In positive psychology self-acceptance, or the deep knowing that we are enough, is a hallmark of mental health. This allows us to look past our limitations and have a lens of self-compassion. The strengths of zest and hope are significantly correlated with self-acceptance (Harzer, 2016).

The Importance of Gratitude

Gratitude is one of the most researched areas of positive psychology. Grateful individuals report higher positive mood, optimism, life satisfaction, vitality, religiousness, spirituality, and less depression and envy than less grateful individuals. Grateful people also tend to be more helpful, supportive, forgiving, empathic, and agreeable (McCullough et al., 2002).

Other studies have found health benefits to the strength of gratitude. "Counting your blessings" was linked to fewer physical symptoms, more optimistic life appraisals, more time exercising, improved well-being, and optimal functioning (Emmons & McCullough, 2003).

Research on Music, Art, and the Humanities

How do character strengths impact music, art, and the humanities? Musicians scored significantly higher on self-regulation and appreciation of beauty and excellence than non-musicians. They also scored lower than amateur and non-musicians on teamwork, fairness, and leadership (Güsewell & Ruch, 2015).

Humanities hold central the appreciation of beauty and excellence. There was research into the importance of this strength within the humanities and the importance of role models (Ruch & Gander, 2020).

The Six Virtue Categories

The 24 strengths are grouped into six virtue categories (Table 1.1). These categories act as a connective tissue for types of strengths and group them according to their function. The groups contain varying numbers of strengths and are classified as such: courage (emotional strengths), temperance (protective strengths), wisdom (cognitive strengths), transcendence (spiritual strengths), humanity (social strengths), and justice (community strengths) ("Character strengths and virtues: A handbook and classification," n.d.).

Table 1.1 The Six Virtue Categories

				Strengths			
Virtue Category	**Courage**	Bravery	Zest	Honesty	Perseverance		
	Temperance	Self-regulation	Prudence	Humility	Forgiveness		
	Wisdom	Love of Learning	Curiosity	Creativity	Perspective	Judgment	
	Transcendence	Hope	Humor	Appreciation of Beauty and Excellence	Spirituality	Gratitude	
	Humanity	Love	Kindness	Social Intelligence			
	Justice	Fairness	Teamwork	Leadership			

We will delve deeply into these virtue categories in the Second Set. There you will juxtapose them with specific music professions and case studies from the field of music. This will help you apply and integrate character strengths into what you already know about your profession and actual people and situations.

The Aware-Explore-Apply Model

Everything about the VIA model is strengths-based (asset-based), focused on individual strengths (personal, social, community), and not deficits (what is wrong with us). Strengths-based practice is holistic and multidisciplinary and works to promote individual well-being. The main advantage of a strengths-based approach is it provides a lens for people to see themselves "at their best" and to see and own their own value. By capitalizing on strengths rather than negative qualities, we move into an open mindset where we can grow and increase well-being.

The aware-explore-apply (AEA) model is the common process by which you learn to work with your character strengths (Niemiec, *A simple model for working with strengths: Via Institute*, 2019).

In the first phase, you cultivate an **awareness** of your strengths, and you learn the language and definitions of the character strengths. You get used to noticing, thinking, and speaking about your strengths as well as spotting them in other people and situations. This initial phase begins to counteract negativity bias and strength blindness patterns as you come to see your own unique variety of positive qualities more clearly.

In the second phase, you **explore** these strengths. This includes examining how you have used them in the past, how you are currently using them, and any planned future use. It's important to use your character strengths both with problems and in good times – for better or for worse! In this phase, you are exploring the possibilities of your signature strengths and ways they can help you thrive. This phase also includes journaling and interactive discussions.

The last phase, **apply**, is referred to as the action phase. Here, you gather and integrate your learnings and insights toward outward action. This is where you set your goals and plan for how you will continue to use your strengths. The apply phase can be creative and fun! Also, think of benchmarks for yourself – What criteria will let you know you have improved? Look toward character strength exemplars in your life, career, movies, and books (Niemiec & Wedding, 2008). Role modeling and the power of observational learning is particularly useful in the action phase (Bandura, 1986). Surprisingly, Niemiec (2009) reported that fewer than one-third of individuals have a meaningful understanding of their strengths!

Strength Spotting – Name, Explain, Affirm, or Appreciate

Strength spotting is a lot of fun! It's rewarding to watch people's faces light up when you notice and name their strengths that you are seeing in action. It is a staple ongoing process in the practice of character strengths and quickly becomes second nature. This practice is done in the context of telling best-self stories, reviewing personal situations or events, and having interactions with others. There are three steps: name, explain, and affirm or appreciate (Table 1.2).

Table 1.2 Name, Explain, Affirm, or Appreciate

Name	Label the strength you see in operation or hear in a story.
Explain	Describe exactly what you observe. Be specific and say how the strength is showing up.
Affirm or appreciate	Acknowledge how that strength shows up and recognize the value it brings. Express your gratitude.

Here are some examples of strength-spotting responses in action:

"I have watched you persevere and discipline yourself to learn this difficult piece of music, and now I see the joy you feel in accomplishing this."

"Your appreciation of beauty and excellence has really shown up in this Instagram campaign. I love your attention to detail and usage of color."

"Thank you for taking the time to weigh all my contract options so carefully. Your attention to detail has saved me a fortune!"

"The way you took your time with every student showed how you really cared about us. You gave everyone's questions equal time, and you made us all feel important."

"I'm so brave! I was so scared to play that solo, but I did it anyway and I'm so proud!"

Train yourself to always be on the lookout for your own and other peoples' strengths – in other words – **strength spot every day!** Be mindful and specific – notice how these strengths show up in your interactions and conversations with others.

Strength spotting is sometimes called "resource priming" – it warms you up before doing anything – a workshop, a performance, or a difficult interaction. It is best to use it throughout all three phases of the AEA model of learning about character strengths. The quintessential exercise in working with character strengths is the suggestion to use a signature strength in a new and unique way (Linley, 2008).

Spotting Strengths in Yourself

After you take the VIA Survey, read over your top strengths, and ask yourself, "Does this fit and sound like me?" Exploring the survey results is the best way to begin having a greater awareness of the strengths in yourself. This is the awareness phase.

Next, be specific: "How and where do these specific strengths show up in my life?" Find a specific example and say it out loud. Use your everyday life stories – your strengths are in the details, and nothing is too big or too small – it all matters. Think about one of your happiest moments recently. What character strengths were present?

Give yourself the gift of spending time to learn and spot your own character strengths. This will develop "**strengths fluency**." This secret superpower will light up your life and the lives of others you share it with. You will never be the same again!

Spotting Strengths in Others

It is far easier to spot strengths in others versus spotting strengths in ourselves. Character strengths are social qualities and seeing them in others will help you make authentic connections. Also, when you validate other people's strengths, they feel seen and special. All musicians can remember that one teacher or person who first spotted their musical inclinations or talents. And those in the music business field will light up when you ask them to regale when they discovered their passion for their work or realized they were adept at their job. This may have taken the form of recognition or self-awareness of being "in flow." Take a moment to personally reflect. Those types of moments become integral to the life you are living today, and the journey of professional decisions you made along the way. Times when others strength-spotted you may have brought you to this present moment!

When you name and express others' strengths as they show up in everyday interactions, you are developing a common language of strengths. It provides an opportunity to reflect on behavior in a new, positive way. When you intentionally express appreciation for others' strengths, your experience together is deepened. This provides a mirror, allowing the person to not only see their strength through your eyes but also to feel that their strengths matter. It is also important to acknowledge when and what things are going well and what you contributed to making that happen. It's not magical – you have an active part and power in things going well.

Strength Blindness

Strength blindness appears in many different ways – from not seeing our strengths at all, to minimizing them, to a general unawareness or disconnection from your personal identity. Or perhaps you see how you use your strengths but don't connect meaning to their usage. Research shows that two-thirds of people are unaware of their strengths (Linley, 2008). While we can usually see the strengths of others, it is common that we cannot see them in ourselves.

While it might seem counterintuitive to you, oftentimes our strengths are so natural to us – like breathing or walking – we just take them for granted, seeing them as very ordinary instead of extraordinary and of value. We don't smell the flowers right in front of us!

Sometimes we know our strengths but have insecurities and lack the courage or ability to put them into action. The VIA Survey is a tool that can give us language and appreciation to counteract that. The VIA model of aware, explore, and especially apply, helps you develop a practice of using and concretizing all your 24 strengths.

And don't do it alone! Find a partner so you can support each other to persist in actively applying your strengths as this will help you push past your insecurities. There are so many groups and networks in the music professions and activities – take advantage of your cohorts! Together, you will help each other spot and appreciate strengths as you both discover new parts of yourselves or bring to light strengths that have been unseen, held down, or forgotten.

Overuse, Underuse, and the Golden Mean

Another important area of character strength research is about the concepts of strengths overuse, underuse, and optimal use. Optimal use of strengths, *the golden mean of character strengths*, refers to expressing the right combination of strengths, to the right degree, and in the right situation. This is the classical notion of the good life, based on the wisdom of Aristotle, Confucius, and Buddha – that balance between too much and too little. Niemiec refers to this as the "sweet spot" of strengths (Niemiec, 2019). After the first generation of character strength research, the second wave was centered around strength imbalances and using strengths in areas of suffering and conflict.

Exploring over and underuse of strengths and looking for "your sweet spot" is a phase of learning that comes after you are very familiar with your strengths, and it is easy for you to spot them. Here, you examine your usage – when you use some too strongly and others too weakly in each situation. Most commonly, there is a dynamic combination of over and under-usage.

Using your character strengths is like writing a great song or a beautiful piece of music – the elements within are balanced in a way that works. Paying attention to these imbalances can offer amazing insights into some of your stuck or less effective behaviors. After all, it is our struggle with this dialectic tension of the imbalances that produces some of our greatest growth!

So, what does finding the golden mean look like? Let's consider the character strength judgment. This involves weighing all ideas, opinions, and facts to make decisions. If you are high in judgment, your inability to decide might simply be an overuse of this strength. In other words, using it *too* strongly is making decisions more difficult and complicated for you.

Here are some examples of optimal use, overuse, and underuse (Table 1.3).

Table 1.3 Optimal Use, Overuse, and Underuse

Optimal use	Bravery: Standing up for yourself in a conflict situation.
Overuse	Kindness: Constantly feeling used and stepped on.
Underuse	Forgiveness: Burdened by resentments from the past.

You will find examples in this book for each strength using a music professional to exemplify overuse and underuse.

The utility and beauty of the VIA Classification system is that you can explore how other strengths can help temper this overuse and get you unstuck so you can move forward. Understanding where and when your character strengths are out of balance is a powerful aspect of creating a more open growth mindset.

Reframing Conflict with Strengths

Viewing conflict through the eyes of appreciating our different strengths is a game changer! It takes a while to truly understand and immerse yourself in your and others' strengths. You naturally will start with your family, the people you live with, hang out with, or work with. Eventually, you come to see that people are very different and no one has the same constellation of character strengths – and that's a good thing! People really behave according to their beliefs and the ways they see life and the world. Using the language of character strengths, we can have words to appreciate our differences. Then we can view them as ways to extend our vision and provide for richer collaborations. In music, imagine how reframing differences allows for a multitude of proficiencies or proclivities to emerge as powerful tools. Think of how by knowing everyone's strength in a group, you can begin to delegate tasks that feel natural to its members. And you will learn quite a bit about people that you can't get to by way of the normal course of conversations. People tend to light up when they discuss these deeper intrinsic values. Try this in your current social group and see for yourself!

Strengths During Hardships and Adversity

The aim of positive psychology is to bring great awareness to the spectrum of what is right with us and find ways we can improve our happiness, all through science and research. It was a much-needed positive side to the whole scope of mental well-being and traditional psychology, which focused mainly on disorders and getting people out of suffering. When we talk about character strengths, we tend to focus on how they elevate our positive feelings and health. However, it's important to also look at the ways strengths play an enormous part in pulling us through suffering and dark times.

The global COVID-19 pandemic forced us all to mine our strengths. Dr. Ryan Niemiec writes a lot about strengths coming through and shining during hardships. He refers to well-being and adversity as the two wings of a bird, both of which are needed to soar (Niemiec, 2022). Our strengths serve these two important roles: appreciating and growing the positives while also buffering and managing adversities and problems. Niemiec's work on the functional roles of strengths during challenging times has extended our knowledge of the science behind them (Niemiec, 2019). There seems to be a process that shows how strengths are used differently prior to, during, and after adversity. Niemiec describes these six functions:

- The Priming Function
  - Strengths are used to prepare for the experience
- The Mindfulness Function
  - Strengths work synergistically to bring presence to the experience
- The Appreciation Function
  - Strengths are used to express value for what has happened
- The Buffering Function
  - Strengths are used to mitigate and prevent problems
- The Reappraisal Function
  - The experience is summarized through the lens of strengths
- The Resilience Function
  - Strengths help us heal and get back to our lives after the experience

Let's look at this in the context of the way a musician might use their strengths if they were undergoing surgery that would force them to step away from their playing.

- Priming
    - Gratitude for catching the problem before it got worse. Bravery to temper the fear of going into surgery.
- Mindfulness
    - Hope that the surgery will go well. Perspective for taking time to step away from their craft because it will be better in the long run. Self-regulation, knowing the time it will take to heal is a temporary hardship but it's very necessary to deal with the problem.
- Appreciation
    - Forgiving their body for having weakness or injury in the first place. Spirituality for seeing a larger theme and realizing they needed rest and were pushing too hard.
- Buffering
    - Hope for coming out of the injury stronger than beforehand. Humor to deal with the process of rehabilitating with some lightness.
- Reappraisal
    - A renewed love for the act of making music. Self-kindness and a gentler practice and performance regime that is more moderate to the body.
- Resilience
    - Leadership by helping others go through the same surgery or by giving advice to prevent others' injuries. Gratitude for coming out of it okay. A renewed zest in their playing and performance.

These functions of our strengths highlight and enhance opportunities while simultaneously balancing and managing adversities. They can do both things at once to increase our thriving – or in other words, act as both of our wings.

Signature strengths are an important subset of strengths when it comes to hardships and getting through traumatic events. These are your most essential and energizing personal qualities that increase well-being and help you manage adversity.

Not only do our strengths help us survive trauma but they also kickstart our growth post-trauma (Casali et al., 2021). Mass tragedies are found to catalyze our strengths. It was discovered that there was an increase in gratitude, hope, kindness, leadership, love, spirituality, and teamwork following the September 11th attack on the World Trade Center (Peterson & Seligman, 2003). It was also found that the more traumatic events someone experiences, the higher their character strengths scores, except for the strengths of love, hope, and gratitude (Peterson et al., 2008).

When struck with the fears and unknowns of the pandemic, we were all forced to go inside, literally, and figuratively, to explore what was there.

There were endless examples of musicians and artists having to completely change the way they were operating to conform to new restrictions, isolation periods, solitary methods of creation, available equipment and technology, emotional and physical health problems, and tempering the chances of getting sick or making others sick. Think of the many ways musicians had to adapt to keep connected to each other, their audiences, their craft, and their livelihood. Shifts happened in every industry touching music, changing operations, sometimes permanently. Virtual meetings and working from home became normative and education had to quickly elicit a bounty of responses to accomplish their same mission. There were many instances of having to quickly learn techniques and methods to keep everything going during isolation periods and lockdowns. Teachers had to very rapidly learn how to record, mix, and produce generally acoustic ensembles to continue making music at any level. So much of life went online or to streaming platforms. Live music on a feed became an accepted way to keep the music playing for so many creators.

Another wonderful way musicians and artists activated their strengths was in the way they came out to perform, connect, spread joy, and make people feel less alone. Music broke out from balconies and in the streets and at times, played by the clanking of whatever instrument was available, including pots and pans to celebrate the resilience and bravery of first responders, workers, and all people who literally put their lives on the line to help society continue. This was a breathtaking bit of humans flourishing in the midst of one of the worst periods of loss and tragedy in our lifetimes. People have an exponential ability to be creative and a drive to thrive in the darkest of times.

Much of the awe, beauty, creativity, and resultant changes that have emerged during and after the COVID-19 pandemic stage represents the dynamic, transitional, nonlinear shape of resilience. It is the golden nectar that can emerge following a difficulty, problem, loss, or tragedy. This is referred to as post-traumatic growth or the transformational arc (Bryngeirsdottir & Halldorsdottir, 2021). It is a rich period of growth following adversity or a tragedy, including better and closer relationships with others, more openness, and receptivity to new possibilities. There is a greater appreciation of life – it is sweeter. As strengths are forged, there are often enhanced personal and spiritual developments (Tedeschi & Calhoun, 2012).

Character strength interventions are plentiful in the positive psychology world. Interventions describe any action or process done to bring about change in people. They have been widely applied with good results, such as leading to less depression and increased happiness (Gander et al., 2012). People emerge from their suffering, better navigate their symptoms, and manage their stress and problems. These interventions have emerged across various settings and populations. In our setting, working with musicians and music professionals – resilience with adversity becomes key to surviving and thriving in a highly competitive fast-moving industry. Tapping into one's strengths is essential for success, in good times and in tough ones.

One cannot deny the presence of stress in musical professions – whether you are a performer, businessperson, or educator, you will encounter both general and very musically specific stresses in your operations. Relating to this, Dr. Kelly McGonical writes about all of the benefits that stress actually provides – the upside to stress (McGonigal, 2016). Her main point is that it is not the stress itself that is the problem, but rather how your body interprets the stress. If your mindset is one of seeing the opportunity as a challenge, rather than a threat, the body responds differently to anxiety or stress. The reaction is more about excitement, like a priming effect, that gets you ready to be at your best. Mindfully making this shift can radically change your feelings about stress and the ways you can use it for power and energy. Yes, stress can exactly improve your performance!

Finally, we've all heard, just "act as if" you know what you're doing. This comes from a psychological principle literally called the "as if" principle (Wiseman, 2014). It infers that the body and brain don't know the difference between the pretend version and the real one. This has given acting, mindfulness, and the use of the imagination incredible power in practicing success. You might feel awkward at first just copying behavior. In neuroscience lingo, you are laying down a new pathway in your brain. You can see this on a brain scan! That's how learning a new role begins for us all; we just imitate. That's how babies learn as well – by watching and imitating. No creativity is needed as we are learning how to do something initially. The more times you do it, however, the deeper the pathway gets. We learn through repetition. At some point we deviate and develop our own style of doing that same thing. You all know about this functionally – it's how you write music or learn a new song.

Both athletes and musicians use the "as if" principle to their advantage in training mode. Besides real practice, they are taught to mentally run through their programs, events, sets, or songs. This similarity in the training opened the door for a correlation in peak performance habits between athletics and music. Many music curricula adopt sports and athletic books on performing at your best. Whereas this imaginative play-through works for both populations, there are many nuanced layers left out of the picture if you only view the participants of each as performers. This includes the creative impetus to perform and write or make music, the desire to share something deeply personal, vulnerability, uncomfortable emotions that can be used as a means of expression, specific ways of communication including the importance of eye contact, and so much more. Part of the frustration we encountered with the sports-themed books on this subject was all these large discrepancies between playing sports and playing music. This was a huge factor and raison d'être for writing this very book!

And lastly, give yourself "permission to be magnificent" (*Inspirational speaker, consultant & positive psychologist Maria Sirois, 2022*) as well as "permission to be human" (Ben-Shahar, 2013). You have both permissions through this book so you can acknowledge and reach for magnificence with grace and be as large as you want to be!

Warm-Up:
Pilot Study

Warm-up

1. *noun*: The way musicians (also athletes, actors, and other types of performers and competitors) prepare their bodies and voices by way of stretching and playing or singing exercises. This is a crucial step in readying oneself before a musical task.

Background

This book is rooted in the birth of the relationship between the two authors. In 2014, Dr. Nancy Kirsner was studying with Dr. Tal Ben-Shahar to become certified in positive psychology at Kripalu Center in Lenox, Massachusetts. The final project in this program was customized by each student to reflect something that would elevate their positive emotions and overall positivity ratio (Fredrickson, 2019).

For this part of the goal, Dr. Kirsner chose to resume singing and dancing, which she greatly enjoyed when she was younger.

It was at this time that Dr. Kirsner became the voice student of Dr. Raina Murnak. Thus began a successful collaboration whereby each shared and taught the other their passions and skills. Dr. Murnak became very intrigued when Dr. Kirsner was creating a music playlist with a song representing each of the 24 VIA character strengths (www.viacharacter.org). An ongoing excitement and dialog emerged around using the character strengths with music students to create a more positive strength-based approach to counteract performance anxiety, ongoing critical evaluations, and negative self-talk. All these named behaviors are counterproductive to spontaneity, creativity, and joyful performance.

Dr. Murnak's educational beliefs and approach were already steeped in strengths-based teaching, a deep understanding of the emotional and psychological life of performers, and an embedded sense of gratitude for life and people. At that time, she was teaching a graduate-level peak performance class for musicians – singers, and instrumentalists – which was exposing them to concepts about positive psychology, the state of flow, communication, the mind, and the body. The stage was set for a wonderful duet between the two.

Initial Work

The first teaching they did together was for a peak performance class in the spring of 2019. Students took the online VIA Survey and then used the

Segura Strengths mat (https://www.strengthclusters.com/news) to learn about their top five signature strengths. They employed the aware-explore-apply model (Niemiec, *A simple model for working with strengths: Via Institute*, 2019). They shared stories about how these strengths showed up in their lives, and later in their performances. As well, they strength-spotted (Beck, *Strength spotting: Discovering others' superpowers*, 2016) each other's strengths as they saw them in their interactions together both inside and outside of their musical lives.

These early explorations with students revealed great excitement and engagement around the oft-ignored positive feedback elements of musical training, aspects of whole, and well-being. They witnessed students beaming with joy, interest, inspiration, pride, and gratitude as they discussed "what they were strong in" rather than "what was wrong" with them. This generated a lot of intrigue and curiosity in the authors who felt this was worth formally exploring.

Pilot Study

The official pilot study entitled *Character Strengths and Musical Performance* commenced in September 2019. There were three 2-hour in-person interventions and one pre-recorded intervention over a three-month period. The website www.positivevoices.weebly.com both tracked the agenda for each session and served as a collection site for surveys and journal responses.

Purpose of the Pilot Study

The study aimed to introduce music students and educators to positive psychology concepts and the VIA character strengths applied to their musical lives.

At the educational level, the goal was to introduce strength-based language, concepts, and applications to both musical training and a musician's self-talk. The main aim was to offset the heavily critical nature of musical training and the resulting negative, weakness-centered mindset of musicians. The hypothesis was a strength-centered approach to music learning and training would naturally feel more joyful, creative, comfortable, and confident to the participants and ultimately, reduce performance anxiety.

Description of Sample

There were eight participants. All self-identified as singers with varying degrees of aptitude for other instruments or skills (music production, etc.). They were between 19 to 35 years of age. There were seven females and one male consisting of five undergraduate students, one graduate student, and two voice teachers.

Methodology

Participants began by taking five initial surveys:

- The VIA character strengths survey (www.viacharacter.org)
- The Dr. Kristin Neff survey on self-compassion (https://self-compassion. org/self-compassion-test/)
- The University of Pennsylvania Authentic Happiness PERMA survey (https://www.authentichappiness.sas.upenn.edu/testcenter)
- The Psychology Today Anxiety Test survey (https://www.psychologytoday. com/us/tests/health/anxiety-test)
- And a survey dealing with their music making and comfort level (https:// docs.google.com/forms/d/e/1FAIpQLSe4wWk84Ga9IoX-xuNivXkE59Uzo DkfWGoWcbL4fumpfc1VcQ/viewform)

Participants reported and kept a record of their answers. They discovered their top-five "signature" strengths which became the basis of most of the work in the sessions. Everyone met as a group three times and watched a fourth intervention online. All sessions contained both individual and group activities. After each session, participants responded to questions in an online journal. At the conclusion of the pilot study, participants re-took each survey, reported their final scores/results, and filled out a qualitative questionnaire about the felt impact of the study.

Description of Sessions

Session 1: Aware

Prior to this first meeting, all participants took the five surveys. A sociometric activity served as a warm-up and group introduction. Dr. Kirsner introduced a brief background of positive psychology, seminal figures, and the origin and application of character strengths. After this, participants did a "walkabout" on the Segura character strengths mat on their top three strengths. They would walk and stand on top of their strength on the large floor mat and were asked how each of them "showed up for them." The question is designed to be open-ended. Dr. Kirsner would ask deeper secondary questions once they began divulging their responses. Dr. Kirsner discussed the concepts of over and under-use of strengths. At the conclusion, we shared thoughts and experiences amongst the group and participants were directed to write feedback in the first online journal form.

Session 2: Explore

The first activity was to tell a story of a strength that showed up since the last session. Each participant was assigned two "strength-spotters" who kept track of the strengths they heard described in the story. The next activity was a two-minute best-self story told in a dyad. Best self-stories, also called strengths or positive introductions, were pioneered by Seligman (Peterson & P., 2004). This asks you to think of a time when you were at your best and identify the signature strengths that were present. This intervention was designed to increase awareness of personal strengths. Each partner took turns speaking and elaborating on that experience while the other listened. At the end of the story, the listener would select all the strengths they heard in the story using character strength cards (see Merch Booth at the end of this book) and add strengths not covered by the VIA. Participants took turns sharing their stories and then strength-spotting their partners.

After this activity, the researchers conducted a psychodrama vignette in which participants chose whose story they wished to see in action in a group-selected/sociometric way. The chosen protagonist then acted out a recent event that was challenging and called upon their character strengths, assisted by other members of the group personifying their strengths. At the conclusion, we shared thoughts and experiences amongst the group, and they were directed to write in the second online journal form.

Session 3: Character Strengths and Performance

The third session was pre-recorded due to participant scheduling conflicts and holiday travel. Each participant watched two videos online.

In the first video, Dr. Murnak began with a personal story of how the strengths infused, reaffirmed, and energized her musical life as a performer. She then went through her own top-five signature strengths through the lens of a recent performance. She explained the power of actively and mindfully deploying one's strengths into their musical life.

In the second video, Dr. Kirsner used her experience performing in a recent dance recital to exemplify each of her top-five signature strengths. She explained also that more commonly, her "performance" is as a therapist and psychodramatist. Akin to musical improv – these endeavors all require spontaneity and creativity. Psychodrama is a form of group psychotherapy in which patients dramatize their own life situations to achieve insight into personalities, relationships, conflicts, and emotional problems, and to alter faulty behavior patterns (*Psychodramatist, n.d.*). They then were instructed to fill out the third journal online.

Session 4: Apply

Participants greeted each other by sharing a strength of theirs that showed up over their Thanksgiving holiday. They used a psychodramatic technique called *concretization* by choosing a specific object in the room to represent that strength. The object served as a catalyst for more spontaneity in their telling of the story. Next, the group got into dyads and told a best-self musical performance story. Partners listened and strength-spotted each other. They turned these stories into a "three-sentence story," which became the basis for a psychodramatic-action sculpture (Giacomucci, 2021). This was followed by a closing summary and discussion of their personal takeaways. They were then instructed to re-take the five initial surveys, report their final scores, and fill out a qualitative final questionnaire about the impact of the study.

Confounding Variables

Impressions and comments should be taken in the context of the small sample size and the fact that the research did not include the participants' performing or music-making. All information is based on self-report and as such, there are inherent issues of accuracy in self-perception and reporting. Not everyone was able to complete the five surveys as a post marker due to holiday travel and timing.

Summary Impressions

In the style of grounded theory research, the information that was most helpful and revealing were the experiential aspects of the sessions, the honesty of the journal entries, and the revealing nature of session discussions. While basic statistical information was available about the changes on the five different survey scales, it was not as strong as the gestalt of the experiential, verbal, and written journal feedback.

In general, participants reported that they understood and accepted themselves more, which allowed them to be more authentic and self-assured when performing. They disclosed that learning about the definitions of character strengths and ways they showed up in their performance and day-to-day lives gave them a language to talk about themselves from a more internal, positive perspective. They all mentioned how there was an almost total emphasis on what they did wrong in music lessons and training and very little discussion of the elements that they did well and were strong. This feedback reinforced the researchers' initial hypothesis of the effectiveness of teaching music populations the language and application of character strengths and a strength-based approach to music training.

Case Study:
Nicole Weber

Artist • Music School Graduate • Pilot Study Participant

Beginning of the Pilot Study

Perspective • Fairness • Social Intelligence • Appreciation of Beauty & Excellence • Forgiveness

Nicole entered the pilot study in the fall of her senior year. She had been a music major and switched majors to communications. Before the study, Nicole says she was in an unsteady headspace and wasn't sure where she wanted to go artistically. She looks back upon her first VIA Survey results and feels that at that time, they were all big themes in her artistry and personal life. She sees her senior year as a practice of changing her perspective. "My friendships were changing, and I was losing touch with one group of girls. This really felt like a loss. Also, in love, I was figuring out dating and letting go of old feelings and patterns." She truly resonated with the results of the survey, "I was using all five of those character strengths daily in music and life. Even my lyrics were focusing on subjects like perspective, fairness and appreciating beauty." Nicole says the study helped her grow as a musician and artist and expanded her mindset. It allowed her to examine both herself and her art with more compassion.

End of the Pilot Study

Spirituality • Love • Curiosity • Perspective • Fairness

Nicole took a second test at the completion of the study. She sees her survey results as "temperature taking" to get a snapshot of her emotions, struggles, and challenges. Since she felt she had experienced a lot of growth in how she was viewing herself as an individual and an artist, she was curious to see if there were any changes in the VIA survey. To her delight, they did change, with her prior signature strengths reordered within her top eight (her high strengths). She attributes these changes to participating in the study at a crucially important time in her life. She was processing the upcoming changes of graduation and figuring out her path. She was consciously utilizing perspective, fairness, and social intelligence every day as tools through this struggle.

Towards the end of the semester, she began working with spiritual practices and was using her perspective differently – as a tool to examine connections between her artistry and spirituality. Going through the training and learning the language and uses of the strengths empowered her. She felt she had a new superpower to help her grow.

The character strengths began to help her clarify her artistic path. Seeing how spirituality rose to the top really resonated with her artistry and personal growth. Curiosity was a big tool in figuring out her life plans post-graduation. Those themes showed up lyrically in her songs as well. So did love, as she wrote about what love could look like with her ideal partner. She was in the headspace of pondering past relationships and learning to let go. Every time she took the survey, she believes it grounded her, gave her more self-compassion, and provided a screenshot of her life that encapsulated where she was on her growth trajectory.

Looking back, Nicole recognized that that semester was her most challenging. She wasn't a music major anymore but was recording music. There were a lot of new beginnings happening, and she was being exceedingly hard on herself. The study helped her to be less so and value her uniqueness and individuality, especially when it came to her music. She began to see the beauty in her artistry. Instead of conforming to the pressure to retrofit herself into pop music cliches – she allowed herself free self-expression. Nicole credits participating in the study using the VIA survey with this. It gave her the "permission" to express her spirituality in her music and openly talk about it in interviews. "This is me, let me celebrate and revel in that, enjoy, and cherish that instead of critiquing myself. As artists, we are prone to be our own harshest critics."

Current Strengths
Love • Spirituality • Gratitude • Hope • Perspective

After the pilot study was over, she started a gratitude practice in 2020. One of the participants in the first study had gratitude as a signature strength and she really admired it and wanted to cultivate it. "That sounded like a cool one that would feel very grounding." As it turns out, gratitude helped her see the value in her art and remain grateful for what she was creating – even if it wasn't exactly how she thought it would turn out. Being more grateful gave her more grace and freedom which is key to her creativity. When she feels more constricted, she's not shining her light or totally being herself.

She keeps going back to the VIA survey because it gives her a screenshot of her current mentality and a push of encouragement: 'keep doing what you're doing! It's working!' Her gratitude practice grounds her today. Her current top five make perfect sense to her!

Nicole is happy with love and spirituality showing up as her top two strengths. "Those two really describe my musical message." She is looking to cultivate more self-regulation, though she doesn't think it's ever going to show up in her top five. She would just love to see it jump a couple of spots.

Control Room:
Your Brain

Control Room

1.   *noun*: The room in a recording studio   and control and mix levels from a
     where producers and engineers sit   recording.

Introduction

New discoveries about the brain in the last fifty years have changed the world of neuroscience and our understanding of human behavior. Neuroplasticity is considered one of the most important medical discoveries in the last 400 years and it has changed the field of neuroscience. Neuroplasticity means that our brains can grow and change throughout our life span. This wasn't even thought to be possible. This finding shifted the entire mindset in many fields (neuroscience, medicine, education) from a closed to an open growth mindset. It also shifted the landscape of human capacities and aging.

As the cartoonist and author Nick Seluk puts it in his children's books, the brain is kind of a big deal! It is the living computer and experience aggregator that commands everything you do, think, and speak. It allows you to navigate and make sense of the world. There are still a lot of questions and mysteries surrounding the workings of our minds and brains. For our purposes, we have focused on the areas of the brain related to activating your character strengths. To begin, let's get to know some brain basics.

The Triune Brain Model

Developed in the 1960s by American neuroscientist Paul MacLean, the triune brain model is based on the division of the human brain into three distinct regions based on evolutionary brain development: your reptilian brain, the limbic system, and the neocortex. This model has been updated since its first adoption (Steffen et al., 2022).

The reptilian brain is also known as your primal brain, and it is made up of the hindbrain and the medulla. It is responsible for survival, drive, and instinct (Budson, *Don't listen to your lizard brain*, 2017). It is responsible for the most basic survival functions, such as heart rate, breathing, body temperature, and orientation in space. These are the involuntary and automatic consistent control mechanisms of our brain.

The limbic system, which includes the amygdala, hippocampus, cingulate gyrus, and other structures, is the seat of emotions. This system in the brain makes us uniquely human as it controls primary attachment or relationship bonds (Šimić et al., 2021). The amygdala is thought of as the stress region as it is responsible for our fight, flight, or freeze response. If you have ever experienced stage fright, you were dealing with your amygdala's response to detecting danger. When activated, it produces stress hormones and chemicals such as cortisol and adrenaline.

The neocortex or the cerebral cortex (neo-mammalian brain) is the third and last layer of the brain to evolve. Our cerebral cortex is responsible for the higher-level processes of the human brain, such as intelligence and personality, problem-solving, learning, language, memory, reasoning, decision making, emotion, impulse control, and reasoning (*Cerebral cortex: What it is, Function & Location,* n.d.).

Modern Theories of the Brain

Newer research emphasizes more interaction between these parts of the brain in everyday life and especially when under stress. None of these areas are discreet, but instead, work together. This understanding led to a new adaptive brain model looking at interdependent function (Steffen et al., 2022).

Brain Vocabulary and Concepts

As a music professional, we don't expect you to have a deep knowledge of brain function, so this section aims to help familiarize you with key terms and parts that may arise in the sections dealing with individual strengths and the brain. While none of us are trying to become neuroscientists, having an idea of where things live in the brain and how the brain works on a basic level informs our learning at a deeper level. It is also an intellectual booster shot encouraging our practices and forming new habits to understand "how things work" in the brain.

Voluntary and Involuntary Functions

Our brains have both. Voluntary functions include things that we choose to make happen such as reading, taking a walk, or learning new music. Involuntary functions happen automatically, without thinking. These include breathing, sweating, heartbeat, and digesting food.

The Nervous System

The nervous system includes your brain, spinal cord, and nerves. It connects your brain to the rest of your body and controls all the actions and feelings of your body.

Nerves and Neurons

Your brain is made up of billions of these nerves and neurons. They are the cells that send messages between your brain and spinal cord to other parts of your body. In other words, your neurons work together to make things happen for you.

Synapses

Synapses are the places or gaps where two neurons meet and communicate their messages by transmitting impulses. When a message leaves the brain, it travels through long, stringy nerves that are like roads connecting your brain to every part of your body.

Three Main Parts of the Brain

The brain has three main parts: the cerebrum, the cerebellum, and the brain stem.

Cerebrum

The cerebrum or front of the brain is the "gray matter" or the cerebral cortex (which you have learned about in the triune model) surrounding deeper white matter (*Brain Anatomy and How the Brain Works*, 2021). This area of your brain has five lobes that function together but handle different tasks (Ranga.nr, 2021): occipital, parietal, temporal, insula, and frontal lobe.

Occipital Lobe This back part of your brain helps you understand what you see. This includes interpreting visual data like light, color, movement, and depth. Think of the ways this part of your brain helps you when you hear or play music.

Parietal Lobe This sits between your frontal and occipital lobes and deals with sensations like pain, touch, and awareness. This part also helps us interpret our proprioception or the body's place in space. Dance and performing on the field in a marching band uses a great deal of this bodily spatial awareness.

Temporal Lobe This bottom part of your brain controls hearing and memory and resides just behind your ears. As musicians and music fans, we use this lobe quite extensively!

Insula Lobe This is a small region deep within the cerebral cortex. It was one of the last areas to be explored since it was so hidden (*Know your brain: Insula*, n.d.). It is said to be a key in our overall awareness and is involved in pain, love, addiction, and even the taste of food and enjoyment of music!

Frontal Lobe This is the largest lobe and is in the front of the head. It is responsible for elements of our personality, executive ability to make decisions, movement, our sense of smell, and our speech. Expressive performance comes from functions of this lobe. The frontal lobe contains the prefrontal cortex, which is said to be the happiness region of the brain. Increased stimulations in this area release the chemicals and hormones that create happiness or well-being (Hanson, *How to trick your brain for happiness*). These hormones and chemicals include dopamine for pleasure, serotonin for calm, oxytocin for love, and endorphins for exhilaration.

This is also where higher-order thinking and concentration happen. Mindfulness occurs here as well as it controls impulses. This area of the brain continues to develop through the age of 25. Think about the tendency of teenagers and people in their early 20s to take more risks or engage in risky behaviors while this brain area is still developing. It's interesting to note also the relatively young ages of musicians who have achieved success. Their ability to "go for it" despite the odds against them comes from this less risk-averse nature of the young brain.

Cerebellum

The cerebellum means "little brain." It is situated above and behind where your spinal cord connects to your head behind the brain stem. The cerebellum sends and receives messages from the cerebrum. The cerebellum helps your muscles work together so you can be balanced to run, jump, and dance (*Cerebellum: What it is, Function & Anatomy*, n.d.). It is shown that people with damage to their cerebellum have difficulties keeping balance, judging distances, and tapping in rhythm. Imagine how important this "little brain" is for you drummers!

Brain Stem

The brain stem is the lower part of your brain that connects the spinal cord to the cerebrum. It contains the centers that control involuntary functions like breathing, swallowing, and sneezing. Singing utilizes this portion of the brain with its heavy focus on regulating large amounts of breath intake (*Brain Stem Death*, n.d.).

Long- and Short-Term Memory

Your brain is always collecting information and storing it in your memory. A new memory – short-term memory – lasts for about 30 seconds before it is lost (Weinstein, 2017). If you practice and repeat something enough times, and it's important, short-term memory can convert it to long-term memory. Think of the crucial role that practice plays in creating long and short-term musical and muscle memory. Also, think about the way you can internally "play" your favorite song in your head. By repetition, you have created a very detailed map of this music, laid down new brain pathways, and encased it into your long-term memory.

Multiple Intelligence

Multiple intelligence is a theory first posited by Harvard developmental psychologist Howard Gardner (*Howard Gardner's theory on multiple intelligences definition and meaning*, 2022), suggesting human intelligence can be differentiated into eight categories: visual-spatial, verbal-linguistic, musical-rhythmic, logical-mathematical, interpersonal, intrapersonal, naturalistic, and bodily-kinesthetic. This contrasted with other ideas about learning, such as the concept of a single intelligence quotient (IQ).

The idea behind the theory of multiple intelligences is that people learn and acquire information in a variety of different ways. The greater the variety of intelligence you access, the stronger and more effective you will learn. It is suggested that knowing and understanding learning styles is valuable to help yourself, or if you work with others. It may also be useful in deciding career and professional paths.

Perhaps most relevant to our focus is the musical-rhythmic intelligence strand. Gardener defined this as an ability to produce and appreciate rhythm, pitch, and timbre, or appreciation of the forms of musical expressiveness (Mills, 2001). He claimed that musical intelligence was "parallel to linguistic (language) intelligence." Those with high musical intelligence learn well by using rhythm or music, enjoy listening to and/or creating music, enjoy rhythmic poetry, and may study better with music in the background (Kelly, 2019).

Interesting Brain Facts

We Have Unlimited Storage Capacity

It has been found that our brains have no limit to the amount of information we can remember and store. Think of how amazing that is for learning and memorizing music! If we had computer hard drives, that equates to roughly 2.5 million gigabytes (Hayward, 2017)! Our short-term memory is not as impressive and can only hold a fraction of that. It is said that we are only able to remember five to nine pieces of information at a time. Think about the way we break up telephone numbers into smaller groups of threes and fours. This aids in our ability to remember (Weinschenk, 2009).

Neurons That Fire Together, Wire Together

When you learn something new, there are accompanying changes in the brain (Neuroscience News, 2021). Engaging in an activity repeatedly makes it second nature and strengthens those neural pathways. This is neuroplasticity in action!

We have also learned that our brains default to a negativity bias. Neuroscientist Dr. Rick Hanson explains this makes negative experiences more prominent and sticky as if they were 'like Velcro." On the other hand, positive feelings slide off the brain as if it was made from Teflon. He explains, "The brain is good at learning from bad experiences, but bad at learning from good ones" (Walesh, Nix your negativity bias, n.d.). Hanson explains you can consciously rewire your brain or use your mind to change your brain for the better (Hanson et al., 2021). We have the ability to reason with ourselves by "talking to our brain" and refuting distorted brain messages. We also should take in and savor all the good things that happen to us, as they tend to go unnoticed in favor of our personal bad news headlines. Humans are unique in their awareness of their own thinking. This is called metacognition and is associated with the pre-frontal cortex. This also highlights the fact that the brain is not your mind. Your mind is the mental activity that creates your subjective reality (Hanson, *Your wonderful brain*). When you think or feel something over and over, you are repeating a pattern of brain activity. These repetitions change actual neural structure and function. You can rewire and change your brain!

Therefore, in life and in music, it's so important to reinforce and review all the positive things that happen to you, not just the negative. It is so easy to re-traumatize yourself with your worst memories of stage fright, failures, and other negative experiences. Keeping a journal of your "wins" helps to offset this negativity bias. Also, using this book to really celebrate and reflect on your strengths can be amazingly empowering!

Hydrate, Hydrate, Hydrate!

Singers have a rather coarse saying, "Pee clear, sing clean." They understand the importance of hydration to the tissues of the body and especially the vocal cords and surrounding areas. Our brains are made of 73% water! And it only takes 2% of dehydration to have an impact on your attention, memory, and other cognitive skills. It was found that exercising and sweating for 90 minutes can temporarily shrink the brain for as much as one year of aging (Alban, 2022)! All this to say, drink up for your body, your instrument, and your brain.

Brains Are Always Running

Our brains are always on, keeping our bodily needs met to keep us alive. They are constantly ready to protect us and tend to dangers. Most of this happens faster than we can be aware of, while the slower events (thoughts and feelings) are more perceivable (Hanson, 2022).

Big Social Brains

There is a hypothesis that says social intelligence is the reason human brains are much bigger than an animal our size needs. Larger brains have developed to keep track of all the complicated social relationships we have (Wayman, 2011). Music is also a uniquely human activity strongly linked to human social contact (Schulkin & Raglan, 2014). Thank your big brain the next time you make music with others.

Tips for Success!

Now that you have a primer on the parts of the brain that you will be reading about through the strengths chapter, let's focus on ways you can best go about adopting the knowledge and practices of this book.

"Making it stick" refers to the science of successful learning, or more simply, what does it take to really learn? This comes from an area called cognitive psychology. In short, there are ways that optimize your learning and retention, rooted in research and neuroscience. Here are a few tips:

1. Step away from concepts and really take time to absorb them. There are spaces in this book to journal – be mindful and take your time! This is about you after all; there are no tests. There is only growth and personal enrichment as a goal.
2. Review the material you have learned – can you summarize it for yourself? Can you teach it to a friend? This helps it stick!
3. Look for as many opportunities as you can, no matter how small, to practice your strengths in daily life.
4. Take note that sometimes when using "new" strategies, they can feel time-consuming and confusing. Stay with it – the rewards will be worth it!

Opening Act Practice Room

Identify where and how your signature strengths are showing up in your life.

You are encouraged to do this activity socially with friends and family. Sometimes we don't see our highest strengths because we assume they are just common to everyone and therefore not particularly special. Our friends and family can help us view the strengths we are "blind" to because they are so natural and normalized to us.

_____
_____
_____
_____
_____
_____
_____
_____
_____
_____

References

Alban, P. (2022, September 21). *72 amazing human brain facts (based on the latest science)*. Be Brain Fit. Retrieved January 9, 2023, from https://bebrainfit.com/human-brain-facts/

Bandura, A., & National Inst of Mental Health (1986). *Social foundations of thought and action: A social cognitive theory*. Prentice-Hall, Inc.

Beck, J. (2016, June 5). *Strength spotting: Discovering others' superpowers*. Medium. Retrieved January 7, 2023, from https://medium.com/@jasonebeck/strength-spotting-discovering-others-superpowers-5ef841aa79fd

Brain Anatomy and How the Brain Works | Johns Hopkins Medicine (2021, July 14). *Brain anatomy and how the brain works*. Retrieved January 9, 2023, from https://www.hopkinsmedicine.org/health/conditions-and-diseases/anatomy-of-the-brain

Illnesses & conditions | NHS inform (n.d.). *Brain stem death*. Retrieved January 9, 2023, from https://www.nhsinform.scot/illnesses-and-conditions/brain-nerves-and-spinal-cord/brain-stem-death#:~:text=The%20brain%20stem%20is%20the,breathing

Bryngeirsdottir, H. S., & Halldorsdottir, S. (2021). The challenging journey from trauma to post-traumatic growth: Lived experiences of facilitating and hindering factors. *Scandinavian Journal of Caring Sciences, 36*(3), 752–768. 10.1111/scs.13037

Budson, A. E. (2017). *Don't listen to your lizard brain*. Psychology Today. Retrieved January 9, 2023, from https://www.psychologytoday.com/us/blog/managing-your-memory/201712/don-t-listen-your-lizard-brain#:~:text=The%20reptilian%20brain%2C%20composed%20of,it%20or%20riding%20a%20bike).

Casali, N., Feraco, T., & Meneghetti, C. (2021). Character strengths sustain mental health and post-traumatic growth during the COVID-19 pandemic. A longitudinal analysis. *Psychology & Health, 37*(12), 1663–1679. 10.1080/08870446.2021.1952587

Cleveland Clinic (n.d.). *Cerebellum: What it is, Function & Anatomy*. Retrieved January 9, 2023, from https://my.clevelandclinic.org/health/body/23418-cerebellum#:~:text=Your%20cerebellum%20is%20a%20part,to%20coordinate%20your%20muscle%20movements.

Cleveland Clinic (n.d.). *Cerebral cortex: What it is, function & location*. Retrieved January 9, 2023, from https://my.clevelandclinic.org/health/articles/23073-cerebral-cortex#:~:text=Each%20of%20these%20lobes%20is,%2C%20emotion%2C%20intelligence%20and%20personality.

VIA Institute on Character (n.d.). *Character strengths and virtues: A handbook and classification*. Retrieved January 7, 2023, from https://www.viacharacter.org/character-strengths-and-virtues

Farlex (n.d.). *Psychodramatist*. The Free Dictionary. Retrieved January 7, 2023, from https://medical-dictionary.thefreedictionary.com/psychodramatist#:~:text=(sī′kə%2Ddrä′,whom%20they%20have%20difficult%20interactions.)

Fredrickson, B (2019). *Positivity*. Harmony Books.

Gander, F., Proyer, R. T., Ruch, W., & Wyss, T. (2012). Strength-based positive interventions: Further evidence for their potential in enhancing well-being and alleviating depression. *Journal of Happiness Studies, 14*(4), 1241–1259. 10.1007/s1 0902-012-9380-0

Giacomucci, S. (2021). Essentials of psychodrama practice. In: *Social work, sociometry, and psychodrama. Psychodrama in counselling, coaching and education*, vol 1. Singapore: Springer. 10.1007/978-981-33-6342-7_13

Güsewell, A. & Ruch, W. (2015). *Character strengths profiles of musicians handbook of methods in positive psychology* (pp. 292–305). New York: Oxford.

Hanson, R. (2022, October 7). *The science of positive brain change*. The science of positive brain change. Retrieved January 9, 2023, from https://www.rickhanson. net/the-science-of-positive-brain-change/

Hanson, R. (n.d.). *Your wonderful brain – Amazon web services*. Your wonderful brain. Retrieved January 9, 2023, from https://s3-us-west-1.amazonaws.com/fwb-media. rickhanson.net/PDFfiles/YourWonderfulBrain.pdf

Hanson, R. R. (n.d.). *How to trick your brain for happiness*. Greater Good. Retrieved January 9, 2023, from https://greatergood.berkeley.edu/article/item/how_to_trick_ your_brain_for_happiness

Hanson, R., Shapiro, S., Hutton-Thamm, E., Hagerty, M. R., & Sullivan, K. P. (2021). Learning to learn from positive experiences. *The Journal of Positive Psychology*, 1–12. 10.1080/17439760.2021.2006759

Harzer, C. (2016). The eudaimonics of human strengths: The relations between character strengths and well-being. Handbook of Eudaimonic Well-Being. 85–92. https://doi.org/10.1007/978-3-319-42445-3_20

Hayward, J. (2017, January 17). *Fascinating things about the brain*. ActiveBeat. Retrieved January 9, 2023, from https://www.activebeat.com/your-health/9-fascinating-things-about-the-brain/

Top Hat (2022, November 24). *Howard Gardner's theory on multiple intelligences definition and meaning*. Retrieved January 9, 2023, from https://tophat.com/glossary/m/ multiple-intelligences/

Maria Sirois, Psy. D (2022, March 15). *Inspirational speaker, consultant & positive psychologist*. Retrieved February 14, 2023, from https://www.mariasirois.com/

Kelly, M. (2019, May 23). *Teaching students who have musical intelligence*. ThoughtCo. Retrieved January 9, 2023, from https://www.thoughtco.com/musical-intelligence-profile-8095

@neurochallenged (n.d.). *Know your brain: Insula*. Retrieved January 9, 2023, from https://neuroscientificallychallenged.com/posts/what-is-insula

Linley, A. (2008). *Average to A+: Realising strengths in yourself and others*. CAPP.

Linley, P. A., Nielsen, K. M., Gillett, R., & Biswas-Diener, R. (2010). Using signature strengths in pursuit of goals: Effects on goal progress, need satisfaction, and well-being, and implications for coaching psychologists. *International Coaching Psychology Review*, 5(1), 6–15.

Lounsbury, J. W., Fisher, L. A., Levy, J. J., & Welsh, D. P. (2009). Investigation of character strengths in relation to the academic success of college students. *Individual Differences Research*, 7(1), 52–69.

Malouff, J. (2018). The impact of Signature character strengths interventions: A meta-analysis. *Journal of Happiness Studies*, 20(4), 1179–1196. https://doi.org/10.1007/s1 0902-018-9990-2

McCullough, M. E., Emmons, R. A., & Tsang, J. (2002). The grateful disposition: A Meta-analysis. *Journal of Happiness Studies*, 20. 10.1007/s10902-018-9990-2

McGonigal, K. (2016). *The upside of stress: Why stress is good for you, and how to get good at it*. Avery.

Mills, S. W. (2001, September). *The role of Musical Intelligence in a multiple intelligences focused elementary school*. Retrieved January 9, 2023, from https://libres.uncg.edu/ir/ asu/f/Mills_Susan_2001_The_Role_of_Musical_Intelligence.pdf

Neuroscience News (2021, December 23). *How neurons that wire together fire together*. Neuroscience News. Retrieved January 9, 2023, from https://neurosciencenews. com/wire-fire-neurons-19835/

Niemiec, R. M. (2013). *Via character strengths: Research and practice (the first 10 years).* SpringerLink. Retrieved January 7, 2023, from https://link.springer.com/chapter/10.1007/978-94-007-4611-4_2

Niemiec, R. M. (2019, May 16). *A simple model for working with strengths: Via Institute.* VIA Institute on Character. Retrieved January 7, 2023, from https://www.viacharacter.org/topics/articles/a-simple-model-for-working-with-strengths

Niemiec, R. M. (2019). Finding the golden mean: The overuse, underuse, and optimal use of character strengths. *Counselling Psychology Quarterly, 32*(3-4), 453–471. 10.1080/09515070.2019.1617674

Niemiec, R. M. (2019). Six functions of character strengths for thriving at times of adversity and opportunity: A theoretical perspective. *Applied Research in Quality of Life, 15*(2), 551–572. 10.1007/s11482-018-9692-2

Niemiec, R. M., & Wedding, D. (2008). *Positive psychology at the movies: Using films to build virtues and character strengths.* Hogrefe Publishing.

Niemiec, R. (2022, September 20). *Well-being and adversity: Two wings to soar: October 2022: Volume 3, issue 5: Via Institute.* VIA Institute on Character. Retrieved February 13, 2023, from https://www.viacharacter.org/news/a-character-strengths-briefing-from-dr-ryan-niemic-october-2022

Park, N., & Peterson, C. (2007). Methodological issues in positive psychology and the assessment of character strengths. In A. D., Ong & van Dulmen, M. H. M. (Eds.), *Oxford handbook of methods in positive psychology,* (pp. 292–305). Oxford University Press.

Park, N., & Peterson, C. (2008). Positive psychology and character strengths: Application to Strengths-based school counseling. *Professional School Counseling, 12*(2), 85–92. https://doi.org/10.5330/PSC.n.2010-12.85

Park, N., & Peterson, C. (2009). Character strengths: Research and practice. Journal of College and Character, 10(4), https://doi.org/10.5330/PSC.n.2010-12.85

Park, N., Peterson, C., & Seligman, M. E. P. (2004). Strengths of character and well-being. *Journal of Social & Clinical Psychology, 23,* 603–619.

Peterson, C., Park, N., Pole, N., D'Andrea, W., & Seligman, M. E. (2008). Strengths of character and posttraumatic growth. *Journal of Traumatic Stress, 21*(2), 214–217. 10.1002/jts.20332

Peterson, C., & Seligman, M. E. P. (2003). Character strengths before and after September 11. *Psychological Science, 14*(4), 381–384. 10.1111/1467-9280.24482

Peterson, C., & Seligman, M. E. P. (2004). *Character strengths and virtues: A handbook and classification.* American Psychological Association; Oxford University Press.

Peterson, T. D., & Peterson, E. W. (2008). Stemming the tide of law student depression: Press. *Psychology, 82,* 112–127.

Ranga.nr (2021, July 22). *The important brain parts and functions: A complete guide.* Study Read. Retrieved January 9, 2023, from https://www.studyread.com/brain-parts-and-functions/

relationships between students' character strengths, school-related affect, and

Ruch, W., & Gander, F. (2020). Ruch, W., & Gander, F. (2022). Character and virtues in the arts and humanities. The Oxford Handbook of the Positive Humanities, 166–177. https://doi.org/10.1093/oxfordhb/9780190064570.013.4.

Sahaja Online (2019, December 3). *Developing character strengths.* Retrieved January 7, 2023, from https://sahajaonline.com/character-personality/character/character-evidence/

Schulkin, J., & Raglan, G. B. (2014, September 17). *The evolution of music and human social capability.* Frontiers in neuroscience. Retrieved January 9, 2023, from https://www.ncbi.nlm.nih.gov/pmc/articles/PMC4166316/

Šimić, G., Tkalčić, M., Vukić, V., Mulc, D., Španić, E., Šagud, M., Olucha-Bordonau, F. E., Vukšić, M., & Hof, P. R. (2021). Understanding emotions: Origins and roles of the amygdala. *Biomolecules, 11*(6), 823. 10.3390/biom11060823

Steffen, P. R., Hedges, D., & Matheson, R. (2022, March 11). *The brain is adaptive not triune: How the brain responds to threat, challenge, and change.* Frontiers. Retrieved January 9, 2023, from https://www.frontiersin.org/articles/10.3389/fpsyt.2022.802606/full

Tay, L., & Pawelski, J. O. (2021). *The Oxford Handbook of the Positive Humanities.* Oxford University Press.

Tedeschi, R., & Calhoun, L. G. (2012). *Trauma and transformation: Growing in the aftermath of suffering.* Sage Publications.

VIA Institute on Character (n.d.). *Via Inventory of Strengths (via-IS).* VIA Institute on Character. Retrieved January 7, 2023, from https://www.viacharacter.org/researchers/assessments/via-is

Wayman, E. R. (2011, October 31). *Humans evolved big brains to be social?* Smithsonian.com. Retrieved January 9, 2023, from https://www.smithsonianmag.com/science-nature/humans-evolved-big-brains-to-be-social-122425811/

Weber, M., Wagner, L., & Ruch, W. (2016). Positive feelings at school: On the relationships between students' character strengths, school-related affect, and school functioning. *Journal of Happiness Studies, 17,* 341–355. 10.1007/s10902-014-9597-1

Weinschenk, S. (2009, October 28). *100 things you should know about people: #3 – you can only remember 3 to 4 things at a time (the magic number 3 or 4).* The Team W Blog. Retrieved January 9, 2023, from https://www.blog.theteamw.com/2009/10/28/100-things-you-should-know-about-people-3-the-magic-number-3-or-3-to-4/

Weinstein, Y. (2017, April 13). *How long is short-term memory? Shorter than you might think.* How long is short-term memory? Shorter than you might think. Duke Academic Resource Center. Retrieved January 9, 2023, from https://arc.duke.edu/how-long-short-term-memory-shorter-you-might-think

Wiseman, R. (2014). *The as if-principle the radically new approach to changing your life.* Simon & Schuster paperbacks.

YouTube (2013). *Tal Ben-Shahar Permission to be Human.* YouTube. Retrieved February 14, 2023, from https://www.youtube.com/watch?v=1hFyjy9P5lg

CHAPTER 2

First Set

Aware – Your 24 Character Strengths

THE COURAGE STRENGTHS

Bravery

Musical Prelude to Bravery

Bravery is one of those strengths that comes with a partner, its own shadow: fear. There is an implication that fear exists in situations where one must conjure bravery. Music professions and fear have many dances together. In turn, bravery is implicitly woven into this trade in so many ways and contexts. Let's unpack some of these shadowy fears.

First, there is the most obvious example for musicians – fear of performing in front of people which ties into the more commonly felt fear of public speaking. This fear tops the list in surveys over the fear of death itself (*Conquering Stage Fright*, n.d.)! To take this to an extreme, people would rather be the main guest at the funeral than the singer at the pulpit. But why is this? Is this fear of exposure? Of being seen and heard? Of talents failing when they are called upon? Of making mistakes in public? Or is it some ghastly combination of all of this? Non-musicians often watch in wonder and awe as performers get on that stage, look out into the crowd, and deliver amidst all those pressures – it almost seems impossible!

And to make things more complicated, no one ever started playing music because they wanted to be filled with anxiety. No youth picks up a guitar or begins singing in a choir because they want to be terrified somewhere down the line. It begs the question – does fear have a secret shadow itself? Is apprehension a shield protecting the sacred act of making and presenting music? Is this fear a form of deep love?

Fear of public presentation extends beyond the stage, touching all industries from teaching to business to therapeutic professions. Other types

DOI: 10.4324/9781003267607-2

of fears abound in these professions which deal with the high risk/high reward potentiality of the creative arts. Most fluctuating perhaps, the music industry itself is built upon change, trends, forecasting, technology shifts, and skill sets that eventually become fossilized in favor of new methods and markets.

Regardless of the side of the industry, music carries a special weight. We tend to put heavy emotional consequences and emphasis on things that matter the most to us. We care deeply about "not messing up" something important. This desire to succeed concretizes the invisible, opposing threat; a tower of fears that threatens to topple the joy at the core of all music professions. Let's climb the bricks of this tower:

- Fear of creating something designed for others
- Fear of being sub-par
- Fear of judgment
- Fear of failure
- Fear of speaking your truth
- Fear of conformity or the need to fit a mold
- Fear of switching jobs or industries
- Fear of the unknown
- Fear of technology
- Fear of becoming obsolete in the industry
- Fear of being responsible for the success of others
- Fear of physically not being able to perform
- Fear of not spending enough time with loved ones
- Fear of not having control of your own destiny
- Fear of injury

It's important to acknowledge and air all of that. With all that's seemingly at stake, it's a marvel that so many people walk bravely into that night. And yet, they do. The drive to create, champion art, go with one's gut, and fully express oneself through music is as old as humanity itself and has proven to be more powerful than the mountains of doubt. Bravery in music is the unwavering decision to charge through your fear, fueled by deep love and inner truth.

What Is Bravery?

Let's do some myth-busting: brave people do feel fear, but they are able to manage and overcome their fear so that it does not stop them from acting. Bravery is about heading into the unknown, ambiguous, messy middle of things despite the uncertainty and risk. Bravery values your convictions about what is right, and you act and deal with the challenges that arise along the journey. You trust that you will find tools, resources, people, and solutions to deal with anything that comes your way.

Bravery comes in many different forms. It can be expressed both loudly and quietly. It is both the child who bounds onto the stage of their first musical like a lion and the child who cries backstage and then takes lamb-like steps onto that stage despite their fear. Whatever form bravery takes, make no mistake – it is consistent, persistent, and goal-oriented. Bravery varies and is contextual, one can be brave in some areas and roles and not in others.

Bravery builds resilience by creating active coping skills to overcome challenges. It is key to personal development and growth as it requires spontaneous thinking in the moment, as well as action and risk-taking. Brave people are compelled to speak up when things are wrong or unfair. This creates trust with others and often benefits the greater good of the group. Bravery helps create and maintain long-term relationships by building a tolerance for vulnerability. This vulnerability also serves to deepen trust in relationships (Niemiec & McGrath, 2019).

Brave people tend to exhibit prosocial behavior, self-efficacy, and an internal locus of control. They have the capacity to reflect, assess risks, and delay their own gratification. Bravery creates a higher tolerance for ambiguity and uncertainty which is a huge asset in our non-linear world. Brave people are drawn to socially worthy pursuits that create meaning in their lives.

Performing music is a wonderful example of bravery. Performing is a complicated set of people and behaviors, with many moving parts. This is the height of spontaneity, and it is filled with ambiguity and uncertainty! It requires continually stoking your bravery to fuel you through. There are famous musicians who suffer from stage fright and still get themselves onto that stage every night! This is brave. When a rapper speaks truth to power in their lyrics defending the marginalized, that is brave. When an insider speaks out on social media about a business practice that is unjust to publishing rights, especially when it could personally injure their reputation, that is brave. When the many women of the #metoo movement unearth their stories after years of self-shame and pain, that is brave.

Bravery and Your Brain

Researchers are learning more about the way our brains react to danger and threats (Hooper, 2018). Clusters of nerve cells in the brain have been pin-pointed to control either a fearful or a brave response to a perceived threat (Goldman, 2018).

Bravery is the ability to act when faced with fear or a threat. The amygdala is the brain structure responsible for fear response and emotional processing (Schiller, 2019). Both humans and animals process fear in this region of the brain. At the most basic level, our fear and anxiety responses are designed to keep us out of harm's way. There is a theory called the Survival Optimization System (SOS) that attempts to codify the various strategies that we employ to keep ourselves safe against recurring threats (Mobbs et al., 2015).

There are strategies and tactics people use to amplify their courage and down-shift their amygdala activation in the face of perceived dangers. For example, self-regulation, another strength, holds hands with bravery in that it helps you get through scary situations both through self-talk and physical practices. The self-talk aspect reminds you that you have gotten through this before, you will get through it now and survive. Self-regulation behavioral practices, such as meditation, deep breathing, and mindfulness can also help you increase your bravery and face your fears (Notebaert, 2020). These practices are very helpful in overcoming and managing anxieties, stage fright, and fear of public speaking. The power of these practices depends on habitual repetition so that they become hard-wired in your brain, and you can call on them.

Affirmation

"I put my fear and anxiety in my backpack, throw it over the wall, and just do it!"

Quick Tip

Face your fear! It will make you stronger!

Musical Examples of Bravery

- Performing despite your fear
- Tackling unfamiliar music at a concert
- Willingness to improvise or share an original song
- Speaking out through your art about causes you believe in
- Using songs to address unpopular or controversial opinions
- Facing critics
- Researching topics others won't touch
- Standing up for your beliefs in the face of peer and other pressure
- Engaging in a physically challenging performance
- Stepping outside of your comfort zone

Discussion Questions/Self-Reflections

What roles and areas in your life could be enhanced with more bravery? Be specific about how the bravery would help you in both the doing and the outcome. How will your opportunities improve as you learn how to harness and cultivate your bravery to face your fears?

_____
_____
_____
_____
_____
_____
_____

Treble and Bass

Treble • Overuse of Bravery

FOOLHARDINESS: risk-taking without prudence or evaluation

- Music attorney convincing their client they can easily win a risky contract dispute.
- Music therapist pushing their patient/client too far/too fast for their progress.

Bass • Underuse of Bravery

COWARDICE: fearful, easily scared, unable to take even small risks.

- Lyricist afraid to say something important and powerful for fear it will be controversial.
- Manager unwilling to stand up for their artist in the face of more powerful entities at the label.

Lights, Camera, Action!

- Perform a song you're scared to sing.
- Step out of your comfort zone and go after an opportunity that has intimidated you.
- Take one small step toward dealing with a fear you've maintained. Break the fear down into small chunks before you begin so that each step is doable and successful.
- Manage your fears – don't put pressure on yourself to 'solve or cure' your anxiety response. Be gentle with yourself and progress through your fears, one small step at a time. This is how it goes for all of us!

Bravery VIP

Barbara Hannigan • Soprano • Conductor

Bravery does not always look like someone running heroically into a fire. Sometimes the greatest act of bravery is pursuing your own truths or dreams when the world around you shuts its doors, hasn't seen it before, or tries to block you out. And not all barricades are visible.

Barbara Hannigan first took to the stage as an operatic soprano and later as a symphonic conductor. Her embodiment of the music both as a performer and conductor is spoken and written about as elements that "move the audience to tears." She redefined bravery and challenged the traditions within both professional worlds of singing and conducting. As a soprano and actress, she challenges the notions of fragile females by playing many of her roles in a strong way. As a conductor, she broke free of what she felt was the expectation of dressing like a man on the stage and celebrated her bare arms as one of her "most expressive tools." She demonstrates a spacious ability to take risks, both creatively and onstage, and in blazing her career path. As a female in a predominantly male profession, she serves as an exciting, courageous role model to young women who aspire to bring their conducting talents to the worldwide symphonic stage (Bartley, 2022).

References

Bartley, W. (2022, December 6). *Barbara Hannigan – Being the music*. The WholeNote. Retrieved January 27, 2023, from https://www.thewholenote.com/index.php/newsroom/feature-stories/25212-barbara-hannigan-being-the-music

Conquering Stage Fright | Anxiety and Depression. (n.d.). *Conquering Stage Fright*. Retrieved February 2, 2023, from https://adaa.org/understanding-anxiety/social-anxiety-disorder/treatment/conquering-stage-fright

Goldman, B. (2018, May 2). *Scientists find fear, courage switches in brain*. News Center. Retrieved January 27, 2023, from https://med.stanford.edu/news/all-news/2018/05/scientists-find-fear-courage-switches-in-brain.html

Hooper, R. (2018, October 19). *The biology of bravery-and fear*. The Wall Street Journal. Retrieved January 27, 2023, from https://www.wsj.com/articles/the-biology-of-braveryand-fear-1539958293

Hack Your Own Brain (n.d.). Karolien Notebaert: Hack your own brain | TED Talk. Retrieved June 28, 2023, from https://www.ted.com/talks/karolien_notebaert_hack_your_own_brain

Mobbs, Dean, Hagan, Cindy C., Dalgleish, Tim, Silston, Brian, & PrÃ©vost, Charlotte (2015). The ecology of human fear: survival optimization and the nervous system. Frontiers in Neuroscience, 910.3389/fnins.2015.00055.

Niemiec, R. M., & McGrath, R. E. (2019). *The power of character strengths: Appreciate and ignite your positive personality*. VIA Institute on Character.

Schiller, D. (2019, October 19). *Bravery in the brain*. Mount Sinai Health System. Retrieved January 27, 2023, from https://www.mountsinai.org/about/newsroom/2015/bravery-in-the-brain

Honesty

Musical Prelude to Honesty

Honesty's best light is that of the mirror. It is fundamental to our happiness to first make this relationship right – that with ourselves. This is where it starts. This may be what touches us most in art – when someone can make something so beautiful and so universally true, it's undeniable. Lyrics that hit a nerve, music that cuts through into our emotional center, and moments that crystalize exactly the way we are feeling – those golden truths resonate for a long time.

So why is honesty such a hard fight? Why do artists go through versions and versions of themselves to get to what was there all along if they are lucky? Why do we lie to ourselves so much? Or is it that we don't hear our own inner voices? We feel off anytime we find ourselves in jobs, relationships, contracts, or identities that do not align with our honesty. Sometimes we get so deeply committed to the road we are on; it feels impossible to even change lanes. We may even go all the way, only to find a hollow victory. And sometimes we just exit. Only after that choice is made, do we again befriend our intuition. Only when we are safely off the highway can we admit to hearing that nagging whisper attempting to course-correct us to our real path.

The music industry likes to groom, mold, and "improve" artists. Sometimes this up-leveling pays great dividends. At other times, artists feel they are grinning and bearing it until the day they have the power and influence to create the music they "really want to make" or to present the image of themselves that is closest to their core.

Here are some ways that all music professionals deal with honesty in all its forms:

- Personal honesty – speaking your mind in your music, being yourself
- Going after the things that matter to you despite consequences like people thinking less of you or losing status in one job you've invested in
- Honesty as a teacher in telling students what they need to do to improve
- Honesty as a judge giving raw truths
- Honesty in the way you dress or present yourself
- Being yourself and speaking out against things that go against your beliefs
- Making artistic choices that are true to your core, your "why," or your mission.
- Singing the truth in your lyrics
- Honesty about who you are or your identity
- Honesty to get out of relationships or jobs that are no longer serving you

And no one said the truth was easy to hear. Honesty can be abused and turn to harshness or at worst, cruelty. It is important to keep gentleness in your honest talk, especially your self-talk. Like all strengths, honesty works best when tempered with partnering strengths. In this case – love, hope, and prudence. But figuring out this balance is well worth the effort and courage. The true work of art at the end of this journey has always been you.

What Is Honesty?

Billy Joel has a well-known song entitled "Honesty," where he describes how difficult it is to find a truly honest person. Honest people are a breath of fresh air – they speak the truth and what you see and hear is congruent with what you receive from them. In other words, their words and actions match up. This character strength can be hard to deliver and find in others as it is delicately balanced by kind, well-chosen words. Honesty is the alignment of one's thoughts and actions deriving from deep personal convictions. In other words, honest people are self-concordant. Their thoughts and actions accurately represent their values and interests. Honesty also strengthens one's ability to accurately self-assess goals, motivations, and competencies. You are more in touch with your true values and interests, which gives you a feeling of peace and a clear purpose.

Musically speaking, this can show up when someone finally realizes that it is not them, but their mother who always wanted to be in the orchestra. This clears them up to decide they do not indeed have a passion for the violin. In another highly significant way, having the ability to accurately check in with oneself allows for career pivots rooted in self-truths. This may show up as an artist and repertoire (A&R) professional who has come to the end of their own interests in tracking new music. They can honestly recognize their passions have shifted, which frees them up to have less guilt or fear as they move forward to their new destination.

Honest people thrive in intimate relationships as they easily take responsibility for their part in things. Honesty is a gift that deepens connections, and it is instrumental in creating healthy, positive relationships based on trust. Honest people tend to display greater consistency in their actions. This in turn gives them a higher sense of mastery over their lives.

It is true that being honest is difficult as the transparency and vulnerability that fuels it require very high levels of emotional intelligence. Often our culture is not comfortable with expressing emotions, therefore emotional vulnerability is not viewed as a desired quality. Also, it is possible to be "too honest" or "brutally honest" as it is sometimes referred to (overuse of honesty). We have seen this type of blunt honesty in the many singing talent shows where there is often one judge who delivers the hard truths without softening them. The music industry is one that requires mastering the paradoxical qualities of thick skin and artistic sensitivity. This can feel like an emotional tightrope to some on the more creative ends of music. This takes some time to develop with lots of experiences – both successes and failures. At the same time, the business of music requires truth when it comes to putting real money and reputation on the line to sell art. Research has shown that the music industry itself has a mental health crisis. A University of Westminster and MusicTank study revealed that nearly 70% of musicians have experienced depression and/or severe panic attacks.

The delivery and "how" people communicate their honesty with each other really matter. Dr. Chris Peterson's positive psychology thesis entitled "Other people matter" (*Other people matter mindset,* 2021) stresses the importance of delivering honesty with kindness and neutrality. Communication grounded in honesty and openness creates harmony that allows people to feel safe and let their guard down. Safety and feeling respected in a musical relationship are especially important as it promotes the openness that creativity requires. These kinds of communication create a greater tolerance and understanding of differences (Hind, 2022). It also lays the groundwork for true collaboration and creates a reciprocal flow for creativity. This respectful honest communication creates a positive stance for talking and working through conflicts (Rosner, 2019).

Honesty and Your Brain

The area of the brain behind the forehead known as the prefrontal cortex becomes activated with truthful, honest behavior. Those who have suffered damage to this area are more apt to lie without a feeling of guilt (Millholland, 2017). Research on honesty and behavior shows that honesty is positively correlated with feelings of normalcy, self-control, and life satisfaction (Błachnio, 2021).

As well, there are many findings on the benefits of honesty. Honesty is linked to healthy positive relationships because you are viewed as trustworthy, and you take personal responsibility for your own actions (Connors, 2020).

It seems most of the research however focuses on deception and why people lie (Ketteler, 2020). We would benefit from more research in people's motivations for being honest.

Affirmation

"I tell people the truth and I'm sensitive to situations where it may be hurtful."

"I am down to earth and without pretense – what you see is what you get."

Quick Tip

Before acting, connect to your own truth.

Musical Examples of Honesty

- Singing from the heart
- Speaking from a personal place before playing a piece
- Sharing difficult personal details in your lyrics
- Staying true to your musical voice in the face of trends
- Telling peers true feedback even if it's hard to hear
- Studying musical subjects that truly interest you
- Telling your artist their new song is subpar
- Being the person in the band whom people trust
- Advising your client to drop their case because it doesn't stand a chance of winning or will have a hefty emotional cost
- Willingness to be raw or unpolished

Discussion Questions/Self-Reflections

Who is someone you admire because of their honesty? What are some ways you wish to be more honest in your life? Additionally, write about a situation when someone told you something honest about yourself that really hit hard but ended up advancing your growth.

_____
_____
_____
_____
_____
_____

Treble and Bass

Treble • Overuse of Honesty

RIGHTEOUSNESS: know it all, critical without boundaries.

- A music critic who steps over the line and character assassinates the artist in a review of an album.
- A teacher who displays elitism and generalizes modern/current music as "bad" without knowing much about it.

Bass • Underuse of Honesty

PHONINESS: insincere, flippant, on the surface response, shallow.

- A branding specialist who is disingenuous in their marketing of an artist.
- A talent agent who feigns interest to get a client they don't intend to put to work.

Lights, Camera, Action!

- Tell your honest opinion in a meeting today in a way that is neutral and kind to others in the room.
- Write your true feelings in a journal as prose, poetry, or lyrics.
- Honor a promise you made and don't cancel that plan.
- Cancel a plan because you're honoring your need to rest and recharge.
- Honestly name a struggle, bad habit, or vice you have been avoiding and don't want to face. But do so without judgment or editorializing about it. Just the facts!

Honesty VIP

Alison Wonderland • DJ

A quality of honesty is rawness. As with food, something raw is uncooked, unprocessed, and not dressed up for different presentations or dishes. It is what it is. People who exhibit honesty connect to the core of who they are and value being that same person across all situations. Australian EDM producer and DJ, Alison Wonderland is a shining example of honesty. She uses her platform to have open conversations on matters of mental health and often shares her own battles and experiences with her fans (Bain, 2020). She candidly shares her life struggles with her fans in a raw and honest way (SBS2Australia, 2018). In interviews, she stresses doing what you truly love in music and listening to your own authentic truth as an artist (*Alison Wonderland Interview – Coachella 2018*, 2018). She was a classically trained cellist who realized her heart was not fully into it. She bravely stepped into less familiar territory, producing music, which clearly paid dividends as she was the highest-billed DJ at the world's most prestigious festival (Coachella, 2018). She has charted with internationally famous artists including Justin Beiber. Alison's ability to share her truth role models not being ashamed and inspires her fans to unburden themselves from their own secrets. She has also been known to give personal advice and cheer people onward about staying true to themselves via DMs (direct messages) to her fans on social media platforms.

References

Bain, K. (2020, October 1). *20 questions with Alison Wonderland: Australian producer talks advising fans through DMS & Meeting Billy Corgan.* Billboard. Retrieved January 28, 2023, from https://www.billboard.com/music/music-news/alison-wonderland-20-questions-interview-9458416/

Błachnio, A. (n.d.). *Be happy, be honest: The role of self-control, self-beliefs, and satisfaction with life in honest behavior.* Journal of Religion and Health. Retrieved January 27, 2023, from https://pubmed.ncbi.nlm.nih.gov/31797252/

Connors, C. D. (2020, June 4). *Honesty – how it benefits you and others.* Medium. Retrieved January 27, 2023, from https://medium.com/the-mission/honesty-how-it-benefits-you-and-others-ecb3e7fabb9a

Hind, J. E. (2022, December 6). *Why is being open and honest so important?* Compass Resolution. Retrieved January 27, 2023, from https://www.compass-resolution.com/2017/02/27/why-is-being-clear-open-and-honest-in-your-communications-so-important/

Ketteler, J. (2020, September 20). *We need to do more research on Honesty.* Scientific American. Retrieved January 27, 2023, from https://www.scientificamerican.com/article/we-need-to-do-more-research-on-honesty/

Manriquez, M. (2019). *The music industry has a mental health problem – university of the Pacific.* Backstage Pass. Retrieved January 27, 2023, from https://scholarlycommons.pacific.edu/cgi/viewcontent.cgi?article=1045&context=backstage-pass

Millholland, L. (2017, January 15). *Area of the brain responsible for honesty identified.* Collegiate Times. Retrieved January 27, 2023, from http://www.collegiatetimes.com/news/virginia_tech/area-of-the-brain-responsible-for-honesty-identified/article_75efc674-4827-11e4-aadd-001a4bcf6878.html

The Positivity Project (2021, August 31). *Other people matter mindset.* Retrieved January 13, 2023, from https://posproject.org/other-people-mindset/

Rosner, H. (2019, May 10). *Collaborate better.* Kellogg Insight. Retrieved January 27, 2023, from https://insight.kellogg.northwestern.edu/article/collaborate_better

YouTube (2018). *Alison Wonderland Interview – Coachella 2018.* YouTube. Retrieved January 28, 2023, from https://www.youtube.com/watch?v=OiXMcJTYDNM

YouTube (2018). *Alison Wonderland: Escaping your lowest point.* YouTube. Retrieved January 28, 2023, from https://www.youtube.com/watch?v=4pgMQvqqRFc

YouTube (2019). Alison Wonderland Says Sticking to Her Intuition was the Best Thing She Ever Did | Billboard. Retrieved June 28, 2023, from https://youtu.be/5Zz2x2qg38g

Perseverance

Musical Prelude to Perseverance

Music is an enjoyable art beloved by most of humanity. The way one feels about their favorite song, musical, movie soundtrack, opera, or band is an emotionally overwhelming and beautiful connection. All those examples are the result of someone's dogged perseverance with thousands of hours and moving parts making the magnificent whole. Every virtuoso pianist you've ever seen started as a small child and has been at it consistently since. What were you doing at age five and can you imagine if you still spent hours every day doing that same thing? Music-making is a story of perseverance. Every hugely successful band you've known has countless tales of winding roads to that success, oftentimes darkened with the shadows of doubt and the feeling that nothing was going to come of all the work.

The perseverance it takes to make it in the industry is also an act of faith and repetition. There are no guarantees that the effort you put in will result in the outcome of success. This is very different from other fields where work equals promotion and advancement. Music is a wholly different beast that absolutely requires many scaffolded levels of perseverance.

First, there is the level of training on your instrument or voice for years. For some, it could mean delving into artistry very young and working in the professional world, ready for all opportunities that come knocking. For others, it means formal training in colleges, conservatories, and universities for years of multiple degrees. The precision of education is just the beginning of the journey. Then there are endless auditions and interviews in the hopes of attaining a small number of orchestral jobs or teaching posts for their specific instrument or specialty. In the pop music world, training is less widely available so much learning comes from doing, networking, and grinding in hopes of getting a following, a record deal, or some other break. In higher education, getting a terminal degree leads to getting a tenure-track job which leads to years of then further proving that you are a known entity nationally in your field. Read this as publishing, performing, obtaining grants, and getting patents all while maintaining excellence in your day-to-day job. Are you exhausted just by reading that? All of these things take so much perseverance, one may become strength-blind because it is just so normalized in their field.

So, what does it take to go that distance, oftentimes through a fog of the unknown? Hope, discipline, belief in the outcome, and a large amount of imagination, especially when the doors are slammed, the answers are no, and the odds look bleak. For the perseverant, this is all an expected part of the journey that they temper with their own belief in the destination. They see the big picture as worth it and they fixate on the distant star, no matter the lightyears. This is how the lightbulb was invented, the first plane flew, and every musical came to be seen. Perseverance is a religion. It is a strong devotion to your beautiful dream. And only those who persevere may know the sweetness of that dream coming true.

What Is Perseverance?

Perseverant people have some great advantages toward success, life resilience, and life satisfaction. They are resourceful, think outside of the box, and work steadily to achieve their goals. Resiliency is a big part of perseverance and teaching yourself to reframe how you see and bounce back from discouragement, distraction, and obstacles. Perseverant people can recognize, admit, and let go when something isn't working, and shift to trying something new. These adaptive qualities help them in every area of life. Another amazing advantage to perseverance is you enjoy the process and take satisfaction and pleasure in completing things.

Anyone who has released any music understands the benefit of perseverance. There are so many micro-steps in artistry and putting forth art. It can feel completely overwhelming to look at the long road toward success in the music industry. Perseverance is a crucial strength to have or build to navigate this journey. Staying flexible with all your experiences, the twists and turns, the work, and managing expectations is all part of this path. Just as strengths hold hands, so do people on your team. People have complementary and different strengths, all working synergistically toward one main goal. It helps to have people on your team who share this strength of perseverance so that you can support each other with all the ups and downs of the process in music.

It is good to know that perseverance is not an inborn trait – it is a learnable skill based on training and life experiences. Resiliency training is part of it and making a conscious choice to reframe your failures as learning experiences. Harvard psychology professor Dr Tal Ben-Shahar teaches his students to "learn to fail, or fail to learn" (*Learn to fail, or fail to learn*, 2017). Failure is a part of life and perseverance carries you through failures. If you don't try or complete things, there is no way to learn and grow. When you approach failure with perseverance, it's easier to see it as a learning experience. Trying again and again, even when it's both frustrating and risky, is how we develop "grit" – a more contemporary version of perseverance. Failures are the rich soil of growth opportunities. You need the perseverance to reach your goals and most of the time, goals take a lot of work. Perseverance teaches you to focus on task completion and gives you the tools to continue working hard until you succeed (Kishore, 2021).

Perseverance internally builds self-confidence and an internal locus of control. You believe in your own ability to accomplish things effectively. Persistent people are seen as dependable and trustworthy. They are good teammates and friends. Perseverance, especially in music professions, is seen as a very desirable quality to have – whether you are in industry roles, recording/engineering roles, artist roles, or education roles. Knowing that your employees or colleagues are equally committed to results and task completion makes for an efficient and trustworthy work environment.

It is easy to believe the myth of overnight success. There are those cases, especially in the day and age of TikTok, but true success, longevity, and wherewithal are a result of people working toward their goals over a long time period – usually multiple years! Many times, the P.R. surrounding an artist appears that they have emerged from a cloud as a sudden success, but with some digging, you can usually trace their roots to a journey of great perseverance, flexibility, openness, and yes, failures too, as they arose.

Perseverance can also improve your relationships and is linked to better mental health. Maintaining close relationships takes work to deal with conflicts and tensions, rather than avoid them. Perseverance provides traction to stay with it. And a real incentive for you – people with close relationships have shown to even live longer (*Strengthen relationships for longer, Healthier Life*, 2011). People with this character strength are at a lower risk for depression and anxiety (*Perseverance toward life goals can fend off depression, anxiety, panic disorders*, 2019).

Perseverance has a dark side too. Doing something again and again that is not working might not always be a wise choice. You can trap yourself with overuse of perseverance. The cost can outweigh the benefits. You need to examine each situation regarding this. This especially relates to music professions. If you are doing the same thing repeatedly and not getting the results you want, you need to pause and examine your labor to see if there is a better application of your energy. Or if you have outgrown old goals and are just going through the motions. In the music field, one may need to face this and reassess if they are still on the best path for where they are now. There may be a more aligned path on the other side of unrequited effort. It's important not to overuse perseverance in this way and come back to yourself for periodic "reality checks."

Perseverance and Your Brain

The neuroscience of perseverance has taught us that creating a steady flow of dopamine is the basis of creating a habit of perseverance (Bergland, 2011). Dopamine is known as the "feel-good" neurochemical transmitter. It is a chemical that transports information and reinforces the habit of perseverance. Try this by taking a 15-minute fun break (do something just for you that is pleasurable), or by visualizing something you are looking forward to or have done in the past. You can tap into your dopamine reserves on demand by learning how to self-administer a dopamine "reward molecule" every day (*Both Sides Now: Dopamine, the Brain's "Reward" Molecule*, 2016). Then begin to imagine a future project and what you will feel like when you complete it and follow through using perseverance. Focus on how good you will feel when you achieve your goals. Every time you complete a task in your life, you have self-administered a hit of "feel good" dopamine and the perseverance habit will be reinforced.

When it comes to making changes in your life – it is not easy. The science of perseverance teaches us how our beliefs can strengthen or weaken our motivation. Psychologists who study motivation and achievement say struggle and short-term failures don't have to make you want to quit if you adopt the right mindset.

There are two fundamental mindsets, or belief systems, that determine how people learn and respond to setbacks and failure. In one mindset, the closed mindset, you get discouraged and give up on your goal. In the other, an open mindset, you embrace the struggle, learn from setbacks, and keep moving forward – you persevere.

For many, perseverance equates to pain and suffering. You must cultivate and use your strength of perspective to reframe how you think of perseverance and see it as a doorway to satisfaction and pleasure. When you ask people who regularly exercise how they stay motivated to persevere through workouts, most say something about knowing how good it makes them feel by the end or after. This is using the future projection technique – "I know once I get going, I will feel better. And I love how my body feels at the end of my workout."

A good behavioral strategy is to assign completion dates for your goals. The use of specific time boundaries amps up the release of dopamine and enhances the excitement upon completion of goals. Think of the dopamine rush you had after completing the achievement of a recent goal. Remember the thoughts and feelings and savor them all over again! It was probably even more exciting if a team or group was celebrating this goal with you.

Above all else, befriend yourself by being on your own side by being your own cheerleader. Along the way, write somewhere visible every step you accomplish to concretize your movement and progress. Celebrate yourself, "Yes! I did it!" This simple action of affirming and congratulating yourself can harness your reward circuitry and tap your dopamine resources. When you don't consciously acknowledge that you have achieved a goal, dopamine is not released, and you will not reinforce the habit of perseverance.

Affirmation

"Everything I do each day is a chance to activate my reward circuits and get a rush of dopamine."
"I work hard to finish what I start."

Quick Tip

Stick with it! Don't give up – take a breath or break – then go back to it.

Musical Examples of Perseverance

- Continuing to play through a piece despite mistakes
- Working extra hard at learning an instrument despite physical, emotional, or mental barriers
- Continuing to study difficult music theory topics or musical analyses
- Applying for a grant repeatedly after rejections
- Continually auditioning for a reality singing show
- Releasing songs and sending them out to music supervisors despite a lack of response
- Reaching out consistently to musical professionals or your mentors for new opportunities and networking connections
- Practicing for many hours

- Re-writing the song that third time
- Composing the score for a musical
- Writing a book
- Scoring a movie requiring multiple ongoing creative relationships
- Grading a stack of papers
- Pushing through physical pain in a performance

Discussion Questions/Self-Reflections

Think of some experience you have considered a failure and describe at least three pearls of wisdom that emerged from that experience. How have you integrated these and moved them into action? Reframe this into your own personal mantra that will strengthen your perseverance. Make that mantra more succinct and write it below and anywhere else you may want to see it daily.

_____
_____
_____
_____
_____
_____

Treble and Bass

Treble • Overuse of Perseverance
OBSESSIVENESS: fixation, pursuit of unattainable goals, not knowing when to let go, quit, or walk away.

- Singer who is dissatisfied with their competition performance only because someone else received a higher score.
- Orchestral conductor who rehearses their orchestra to the point of exhaustion and they end up lifeless for the performance.

Bass • Underuse of Perseverance
FRAGILITY: easily thrown off your goals or path, too responsive and attached to criticism, allowing others to break your dreams.

- Musicologist giving up publishing an article after one rejection.
- Music theorist unwilling to present at a conference for fear of peer criticism.

Lights, Camera, Action!
- Audition for an opportunity you've been turned down for – keep going!
- Set a timer and complete something that has been sitting on your desk for a while. To set yourself up for success, choose something you think you can handle in 30 minutes.
- Practice that difficult passage until it's easier to master.

- Play a puzzle or word game on your phone and don't give up until you solve it.
- Finish writing that song.
- Set a timer and send out those emails you have been putting off.
- Make a daily checklist of all tasks you want to achieve and reward yourself with a tiny celebration each time you write in that check mark. And remember, tomorrow is another day for another list!

Perseverance VIP

Dolly Parton • Songwriter • Artist

Perseverance is often infused into the children of families who worked tremendously and struggled to make ends meet. The value of hard work is ingrained into children who see their parents do what they need to do to provide. Obstacles and hardships are part of the terrain. Not quitting despite discouragement builds character to get what you need or want. The length and breadth of singer/songwriter/actress/author/entrepreneur Dolly Parton's career speaks to her perseverance (*Dolly Parton*, 2023). Beginning as a songwriter for others, Dolly honed her talents and eventually became a country music performer in her own right, earning her countless industry accolades. As 1 of 12 children born in poverty in Pitman Center, Tennessee, she persevered and worked extremely hard even as a child performer. She eventually wrote songs including "Coat of Many Colors" describing her family's struggles. Through constant hard work, Parton landed a spot on television, was signed as a writer and then an artist, crossed over into more mainstream pop music, penned award-winning songs for herself and others and for film, starred in movies, and opened a theme park on the grounds of her family's meager beginnings in Tennessee called Dollywood. Dolly still works, keeping herself current and relevant by collaborating with young country stars and pop groups. Her perseverance, through different eras and restrictions for women in music and business, truly shines.

References

American Psychological Association (n.d.). *Perseverance toward life goals can fend off depression, anxiety, panic disorders*. American Psychological Association. Retrieved January 28, 2023, from https://www.apa.org/news/press/releases/2019/05/goals-perseverance

Bergland, C. (2011, December 26). *The Neuroscience of Perseverance*. Psychology Today. Retrieved January 28, 2023, from https://www.psychologytoday.com/us/blog/the-athletes-way/201112/the-neuroscience-perseverance

Pennmedicine.org (2016, February 29). *Both Sides Now: Dopamine, the Brain's "Reward" Molecule*. Retrieved January 13, 2023, from https://www.pennmedicine.org/news/news-releases/2016/february/both-sides-now-dopamine-the-br

Kishore, K. (2021, September 2). *What is perseverance?* Harappa. Retrieved January 28, 2023, from https://harappa.education/harappa-diaries/what-is-perseverance/

Power2Improve (2017, May 23). *Learn to fail, or fail to learn*. Retrieved January 28, 2023, from https://www.power2improve.com/en/blog/2015/04/21/learn-to-fail-or-fail-to-learn/

Harvard Health (2011, January 18). *Strengthen relationships for longer, Healthier Life.* Retrieved January 28, 2023, from https://www.health.harvard.edu/healthbeat/strengthen-relationships-for-longer-healthier-life

Wikimedia Foundation (2023, January 27). *Dolly Parton*. Wikipedia. Retrieved January 28, 2023, from https://en.wikipedia.org/wiki/Dolly_Parton

Zest

Musical Prelude to Zest

Zest is the pilot light that burns deeply within every day, connection, and task. It's the fire behind every creative endeavor, be it mobilizing a marching band, engaging in deep research, or producing the next rap hit. Zest is the speech that catalyzes coders to work tirelessly on new music tech products. It's the spark in the stadium that contagiously spreads a fight song. And it's the adrenaline quickening everyone's heartbeat working behind the scenes at a major music awards show.

To live with zest is to eat the energies of renewal for breakfast. You awaken with gratitude for each living moment and immerse yourself in the present with joyous energy. In other words, you woke up like this and plan to spend the rest of the day like this as well. You can spread your light to all whom you meet and greet in your day, raising the energies of others in the room. You are such an important, and perhaps underestimated, stimulating spirit to processes. We have already discussed the huge amount of perseverance needed in this industry. Zest serves as an energetic inspirational leader.

People with zest may not even be aware of how their natural jolt affects the other circuits in the system. Those around them may rely on their brightness and zeal without realizing it also. When someone with high zest isn't involved in a project, it might just seem less fun or more like drudgery. Zestful people have the natural ability to fuse ease and lightness into even the most mundane, regular scenarios that most people would dread. These boring scenes only challenge them to mine deeper and find something to love and get excited about.

Music needs zest and enthusiasm to have a full life cycle. Sometimes that comes from the artist and sometimes even more importantly from droves of fans losing it during the artist's performance, then sharing zealously on their social media platforms, thus encouraging more fervor. The music industry capitalizes on the excitement surrounding releases, tours, merchandise, and appearances of their artists. Advertising agencies spend tons of money on syncs, not because they want to break the bank, but because they know the appeal of beloved songs is so strong, they will have a proximal effect on the product.

Teachers also need to exhibit zest. They stand at the front of many types of rooms, filled with many types of learners, in many types of moods. When a teacher can model a passion for their lesson or subject, they have the power to captivate and fully wrap in their students. When they have a strong authentic fire, they can bet their students will become disciples of lit matches who will pass on that gift of curiosity, love of learning, and general joy.

So light more fires, engage with more people with genuine interest, and lose yourself in the beauty or excitement of music, the moment, your relationships, and your daily tasks. Don't view those as a grind but as a gift – you are lucky enough to be in this exact moment, learning and growing through exactly what you are meant to be shown. Zest is life.

What Is Zest?

Individuals high in zest are like energizer bunnies – they are always full of life and ready to go. They are fully charged from a balanced life of physical and mental activity, and they take care to eat well, sleep, and exercise (*Strengths spotlight #15: Zest*, 2020). Depending upon your own level of zest, being around them can be fun and motivating, or you might feel exhausted just watching them. They embody the spontaneity and creativity of children, and their zest can be contagious, attracting you to want to be around them. In the same way, students can "catch" enthusiasm from their zesty teacher who is passionate about the subject they are sharing. Their positive energy and emotions can spread through the room.

If you are this zesty person, you approach your life with interest, enthusiasm, and positivity. You are very present in the moment; using all your senses, thoughts, and feelings to experience every drop of life, and you appreciate each experience as unique and unfolding. There is a component of mindful awareness in zest.

People high in zest tend to have better relationships as they possess empathy, good communication, and the ability to manage conflict very effectively. They can see the other person's point of view. They know how to build secure connections which bring satisfaction and happiness at work and to their families. Zestful people are less prone to burnout and stress attacks (Chowdhury, 2022). As employees, they are tangible and likable assets to the team or organization. They also exhibit higher endurance and adaptability in service jobs like teaching or medicine. Zest also correlates positively with autonomy and goal attainment (Martela et al., 2017).

Zest is usually present in those who are superfans of music. It is also great when that superfan is themselves in the industry, in roles like A&R (artist and repertoire) or music supervision, or an artist themself. Zest may even be the character strength of your favorite teacher, the one who pushed you to your personal best and rallied with you through your hardships and uphill climbs.

Zest and Your Brain

Research on the benefits of zest show that it is strongly connected with the aspects of happiness including engagement, pleasure, and meaning (Vella-Brodrick et al., 2009). The energy of zest creates a fuller expression of all our abilities and talents. It attracts and draws people in by inspiring and motivating them to get more involved. People with zest tend to view their work as a "calling" and it is fulfilling and meaningful to them.

Rexford Hersey examined employees' feelings about their jobs.

He discovered that one's zest and productivity can change from day to day, depending on other changing physiological, psychological, and sociological factors (Hersey, 1955).

Later studies showed evidence that some people are naturally zestier than others (Chowdhury, 2022). According to scientist Daniel Kahneman, we experience about 20,000 moments each day. Zest is the gift of experiencing each of the moments in "full color" as they are occurring (Rath & Clifton, 2022).

Affirmation

"I see life as one great adventure."
"I approach my relationships and whatever I do with energy and excitement."
"Whatever I do, I'm in!"

Quick Tip

Be alive to the beauty and life energy that surrounds you in every moment!

Musical Examples of Zest

- Encouraging the choir as a section leader through difficult music
- Enthusiastically warming up your voice or instrument
- Eagerly assuming a role in a play or musical and getting into the character
- Getting excited about the whole journey of musical learning
- Volunteering to take on leadership roles
- Bringing excitement and energy to a rehearsal
- Making the auditionees feel comfortable and excited to perform
- Motivating your bandmates to bring their A-game to the gig

Discussion Questions/Self-Reflections

Who brings the most zest to your team, job, or musical organization? What specifically do they do? How does that make you feel? If it is you, how does this impact the group? In what areas of your life would you like more zest? What can you do to increase your zest?

_____
_____
_____
_____
_____
_____

Treble and Bass

Treble • Overuse of Zest

HYPERACTIVITY: your energy levels are way out of sync with your partner or the group and it creates a tension and disconnect.

- A program director who overextends their faculty on multiple initiatives or projects.
- An audio engineer who constantly over-books themselves to the detriment of the quality of their work.

Bass • Underuse of Zest

SEDENTARY: passivity, restraint, people feel your lethargy, which can create resentment as people feel they have to "carry you."

- A songwriter who never finishes a song.
- A music merch salesperson who is bad at networking.

Lights, Camera, Action!

- Create a playlist of songs you love and blast them in your house or car and dance and move!
- Program a concert full of songs that energize and lift people up.
- Go the literal extra mile and travel to see a band or performer you love.
- Take a day to let out your inner child!
- Take time to really play with your pet.
- Write a positive comment for a YouTube video you saw and loved.

Zest VIP

Willi Ninja • Dancer • Choreographer

Zest is the energy that moves mountains, the excitement of starting a day, bringing your A-game, and passion for your role in the play we call life. And there are no small parts for a person with zest! Perhaps there is no better example of zest than dancer, choreographer, musician, and runway model Willi Ninja who was known as the "Grandfather of Vogue." If you have ever watched the show Pose or have emulated the dancers' articulated hands in Madonna's "Vogue" video, you have experienced the fabulous essence of the New York City Drag Ball scene of the 1980s, resplendent with flamboyant and colorful costumes, singing, and dancing. Willi became the main figure of this epoch after appearing in the documentary *Paris Is Burning*. He bravely and zestfully represented the LGBTQ and Black homosexual communities and brought "vogueing" into vogue (*Willi Ninja 2005 Interview & voguing on talk show*, 2019). He was sought after to choreograph and coach models in this new dance form and on poise and walking. Every entrance of Willi's was a grand entrance and he exuded spirit, flair, and confidence with each step, owning and celebrating the clothes, accessories, and space he was in (Herrera, *Willi*

Ninja: Voguing butch queen · challenging gender boundaries: A trans biography project by students of Dr. Catherine Jacquet). He is a true example of someone who rose each day with zest and brought that excitement to the lives of those he graced with his presence.

References

Chowdhury, M. R. (2022, August 21). *Zest in positive psychology: Satisfaction in life and work.* PositivePsychology.com. Retrieved January 28, 2023, from https://positivepsychology.com/zest/

Herrera, A. (n.d.). *Willi Ninja: Voguing butch queen – challenging gender boundaries: A trans biography project by students of dr. Catherine Jacquet, outhistory.* outhistory.org. Retrieved January 28, 2023, from https://outhistory.org/exhibits/show/tgi-bios/willi-ninja

Hersey, R. (1955). *Zest for work—Industry rediscovers the individual.* Harper & Brothers.

Institute of Positive Education (2020, August 31). *Strengths spotlight #15: Zest.* Retrieved January 28, 2023, from https://instituteofpositiveeducation.com/blogs/strengths-spotlight-podcast/strengths-spotlight-15-zest

Rath, T., & Clifton, D. O. (2022, March 24). *The big impact of small interactions.* Gallup.com. Retrieved January 28, 2023, from https://news.gallup.com/businessjournal/12916/big-impact-small-interactions.aspx#:~:text=According%20to%20Nobel%20Prize%2Dwinning,moment%22%20lasts%20a%20few%20seconds

Martela, F., Ryan, R. M., & Steger, M. F. (2017, March 28). *Meaningfulness as satisfaction of autonomy, competence, relatedness, and beneficence: Comparing the four satisfactions and positive affect as predictors of meaning in life – Journal of happiness studies.* SpringerLink. Retrieved January 28, 2023, from https://link.springer.com/article/10.1007/s10902-017-9869-7

Samples. *Social Indicators Research, 90*(2), 165–179. http://www.jstor.org/stable/27734781

Vella-Brodrick, Dianne A., Park, Nansook, & Peterson, Christopher (2009). Three Ways to Be Happy: Pleasure, Engagement, and Meaning—Findings from Australian and US Samples. *Social Indicators Research, 90*, 165–179. 10.1007/s11205-008-9251-6

YouTube (2019). *Willi Ninja 2005 Interview & voguing on talk show.* YouTube. Retrieved January 28, 2023, from https://www.youtube.com/watch?v=PpRG5_VyZKM

THE TEMPERANCE STRENGTHS

Self-Regulation

Musical Prelude to Self-Regulation

Self-regulation is balance. It's knowing and doing what is best for you, not just in the moment you are in, but in service to future moments. As appealing as it may seem, one can't take a daily tour of the chocolate factory. Even if you have the golden ticket.

Making music is an act of self-regulation with perseverance and zest fueling its engine. Youth who engage in music seriously have foregone many social events in favor of lessons, extended practice sessions, and performances. These all may require travel, competition, preparation, and auditions. In other words, self-regulation is the tool employed to temper the spontaneity of being young with the intentionality of having a long-range goal.

Self-regulation is an important part of caring for the body which is our main instrument, no matter the profession or endeavor. It's crucial to learn how to properly care for yourself, especially if it quite literally is your instrument as is the case with singers. Make no mistake, even if your job consists of sitting at a desk, your body's harmony or disharmony will show itself in your work, attention span, physical comfort, and disposition. It's important to dedicate real inquiry and effort to getting your health balance right.

Nutritional regulation plays a grand role in the way your body's chemicals and moods are manufactured. When musicians and professionals are on the road, it is very easy to give in to the convenience of fast food and non-healthy choices. A great example of self-regulatory behavior would be getting up early to get that run or walk in, meditating to center yourself on the day's goals, and having a nutritious breakfast to energize you for the adventure to come.

So many "behind the music" specials are themed with dramatic downward spirals out of self-regulation. It's so easy to over-indulge in all types of things in the industry. At the highest levels, one has access to parties, substances, spending, encounters, fan adoration, and worlds of endless excesses. Meteoric rises to fame can be overwhelming and absolutely challenge one's relationship to their own self-regulation. Tales of people in the arts and sports having everything but losing or spending it all are far too common.

Superstars with careers spanning decades are case studies. They may not have always had a stable sense of self-regulation, but somewhere along the line, they figured out the equilibrium. Many times, it takes tumbling off the scales of balance to teach limits and boundaries. These professionals have figured out a way not only to be internally intact but also to maneuver the sharp turns and constantly changing trends to stay on top of the charts. They have teams that help them make smart and measured choices and they are wise enough to value those people. They listen to sound advice and pivot and morph when needed. The result of this careful management is long-term survival, success, thriving, and growth.

Self-regulation is about playing a long game. It's about learning your own personal limits and stepping back several paces so you can live to fight another day. Emotional regulation is key to this. The ability to take a breath and consider your next decision is the invisible superpower behind the moves you make.

Though it may not seem like the flashiest strength, it is the one that brings it all back to you. It checks in to see if you're alright, if something feels off, and if there's anything it can do to help. Self-regulation is the friend who has your back and keeps your best interests at heart.

What Is Self-Regulation?

There are many different areas of self-regulation: in interpersonal relationships, in exercise, in health and food, and in the world of work. Simply put, this strength is about discipline, and managing your negative emotions, impulses, and excesses. Self-regulation means having the self-control to contain and regulate your words and actions (Cuncic, 2022). Depending on your emotional makeup and how you were raised, self-regulation might be part of how you were brought up or it can be very foreign to you. Self-regulation is an ongoing process and practice, requiring commitment and a willingness to self-reflect; like having a pause button you can push at will. You must train yourself, just like you train the muscles in your body. After a while, the monitoring of your thoughts and actions comes more naturally, and then it becomes a good habit. Self-regulation puts you in charge of your own life.

On the surface, this strength sounds pretty dry. However, this is an unfair assessment. Self-regulation is the complex super-power that allows you to navigate and manage your differing energy states. In all areas of life, the ability to contain and self-regulate will keep you grounded, calm, and focused (*What is self-regulation?*, 2022). Self-regulation is not static or equal in all areas of your life. Emotions, lack of sleep, level of investment, and many other real-life factors can impact it. As with any exercise, strengthening your self-regulation muscles is key to achieving any long-term goal. Also, your behavioral discipline enables you to maintain health throughout your life span. Self-regulation allows you to think more carefully about your actions, make healthier choices, avoid dangerous impulses, and inhibit unwanted thoughts. Sometimes less is more (Heatherton & Wagner, 2011)!

Self-Regulation and Your Brain

The prefrontal cortex is the main player in self-regulation. Its role is to control subcortical regions involved in reward (striatum) and emotion (amygdala) (Hare et al., 2005). Inhibition is a core aspect of self-regulation. This is a process of initiating, adjusting, or stopping thoughts, feelings, or actions to maintain your goals (Kelley et al., 2018).

Affirmation

"I take a breath, think, and then consciously regulate what I say and do."
"I am in control of my emotions, and I don't let them run the show."

Quick Tip

Manage your excesses and vices.
Think before you act.

Musical Examples of Self-Regulation

- Keeping a regular practice routine
- Staying at home, resting, hydrating, and eating well the night before a big show
- Committing yourself to consistent exercise or yoga to help you feel, think, and play better
- Pacing of oneself during long marching band rehearsal
- Setting serious deadlines for your own album release
- Knowing when to stop drinking with the band
- Finding common ground to work with people you don't necessarily like
- Maintaining composure while a label magnate yells in the meeting
- Carefully thinking about how to discipline a poorly behaving class

Discussion Questions/Self-Reflections

In which areas of your musical life could you use more self-regulation? What behaviors could you put in place to achieve this? Are there areas in which you are over-using self-regulation/holding yourself back too much? Write about the benefits of both.

_____

_____

_____

_____

_____

_____

Treble and Bass

Treble • Overuse of Self-Regulation
INHIBITION: overly shy, unable to assert oneself in their task, contracted, narrow and restricted responses.

- DJ who cannot make eye contact with the audience.
- Church choir director who is too shy to ask members to join the choir.

Bass • Underuse of Self-Regulation
SELF-INDULGENCE: going too far, self-centeredness, "it's all about me."

- Professor taking all performance opportunities that could go to their students.
- Producer who only values their own sound and not the vision of the artist they are working with.

Lights, Camera, Action!

- Create a warm-up priming ritual that helps you perform at your best and implement it each time you get out there – on the stage, on the air, in the studio, or in the boardroom.

- Increase your daily water intake. Take care of the body and brain that gives you so much talent and joy.
- When you are anxious, irritated, or feeling reactive – notice it and take a mindful pause. Take at least six deep breaths and then move into your normal breathing pattern. If you can, move out of the physical space to help shift the energy.
- Try sensate focus for finding the present. Breathe more intentionally and slightly slower. Look around in your setting and notice what and who you see; what you hear, smell, and can touch. Come into your senses and get out of your head.

Self-Regulation VIP
Liz Caplan • Performance Coach

Music training requires an abundance of self-regulation. Each day you pick up your instrument, do a vocal warm-up, or mentally run through that movement, you are activating your self-regulation. And just like with music, practice makes perfect for energizing this strength. Not only did Liz Caplan get to where she is now by self-regulation, but she channels and models that strength for her clients and students to get where they want to go as well (Hetrick & Gioia, 2014). As a leading Broadway vocal coach and teacher, Caplan lives in a state of maintenance and self-care (*Liz Caplan • Maestra*, 2020). She speaks of the importance of taking care of the whole body systemically to keep the voice healthy. She also believes in ritualistically warming up to get the most out of your voice and keep it flexible and healthy (Playbill Video, 2018). Liz tracks her dietary intake, amount of sleep, and allergies – all these things need to be regulated for her to perform at her peak. Clients who have worked with her rely on her adherence to her habits and behaviors around self-regulation to ensure they are treating their own voices well.

References

Cuncic, A. (2022, January 27). *How to develop and use self-regulation in your life.* Verywell Mind. Retrieved January 28, 2023, from https://www.verywellmind.com/how-you-can-practice-self-regulation-4163536

Hare, T. A., Tottenham, N., Davidson, M. C., Glover, G. H., & Casey, B. J. (2005, March 15). *Contributions of amygdala and striatal activity in emotion regulation.* Biological Psychiatry. Retrieved January 28, 2023, from https://pubmed.ncbi.nlm.nih.gov/15780849/

Heatherton, T. F., & Wagner, D. D. (2011, March). *Cognitive neuroscience of self-regulation failure.* Trends in Cognitive Sciences. Retrieved January 28, 2023, from https://www.ncbi.nlm.nih.gov/pmc/articles/PMC3062191/

Hetrick, A., & Gioia, M. (2014, June 9). *Meet Liz Caplan, the vocal coach behind the tony award-winning stars of Hedwig, Aladdin and Les Miz.* Playbill. Retrieved January 28, 2023, from https://www.playbill.com/article/meet-liz-caplan-the-vocal-coach-behind-the-tony-award-winning-stars-of-hedwig-aladdin-and-les-miz-com-322262

In The Studio: Vocal Coach Liz Caplan (n.d.). Facebook. Retrieved June 28, 2023, from https://www.facebook.com/playbill/videos/in-the-studio-vocal-coach-liz-caplan/10155246180295418/

Kelley, N. J., Gallucci, A., Riva, P., Romero Lauro, L. J., & Schmeichel, B. J. (2018, December 21). *Stimulating self-regulation: A review of non-invasive brain stimulation studies of goal-directed behavior.* Frontiers. Retrieved January 28, 2023, from https://www.frontiersin.org/articles/10.3389/fnbeh.2018.00337/full#:~:text=Neuroscientific%20research%20has%20revealed%20that,emotion%20(e.g.%2C%20amygdala)

Maestra. (2020, March 28). *Liz Caplan • Maestra.* Retrieved January 28, 2023, from https://maestramusic.org/profile/liz-caplan/

Yourtherapysource (2022, June 6). *What is self-regulation?* Your Therapy Source. Retrieved January 28, 2023, from https://www.yourtherapysource.com/blog1/2020/01/19/what-is-self-regulation-2/

YouTube (2018). *Broadway's Elite Vocal Coach, Liz Caplan, Shows Us Her Secret Warm-Up.* YouTube. Retrieved January 28, 2023, from https://www.youtube.com/watch?v=D4tW6w51yAI

Humility

Musical Prelude to Humility

There are so many spotlights to seek in the field of music. Within the performance spectrum, the competitive academic landscape, and the fast-paced business terrain, one is trained to announce themselves, wave their own flags, grab the brass ring, and remind the world why they are special. And when we add the barrage of the vanity press that is social media, humility seems to be largely bred out of us. This is why it is such a breath of fresh air to witness someone do their part with silent strength and grace, without the need to call themselves out. Humility is a quiet confidence in the value of one's part in the play while giving props to all the supporting cast and crew.

People who display humility consciously cast aside the huge pressure to be the one-man show. They accept their contribution as only part of the reason something succeeds or fails for that matter. There is a Zen-like peace that exists around the truly humble and it also feels like a great gift when others get to give that person the praise they deserve. They operate from a force outside of the ego which makes people feel safe. They can be great collaborators who give space to their teammates. Also, because it doesn't diminish them from that internal place, they freely admit to their mistakes and shortcomings and have no problem asking questions when they don't know something. What a refreshing energy in any endeavor. Their grounding nature can help anchor those around them.

The industry can feel like a rollercoaster – lots of highs and lows. Sometimes you're at the top of the park, seeing it all from a high vantage point only to next find yourself careening back to the bottom at a high speed. It is important to keep humility close on your ascents, realizing that you may be one turn away from a descent, and the cycle may repeat. You can have a band opening for you and then six months later they might be headlining venues three times the size of yours. If you didn't treat them with respect, they will remember that. It's important to be kind and empathetic to people on all legs of the journey. Remember, you may walk in everyone's shoes at some point.

Showing humility does not diminish your own worth. There are surely situations where you must self-promote, break through ceilings, and toot your own horn to make people aware of your contributions. However, when the ticker tape from the parade settles, it is of great importance to be mindful of the village which is also a part of your success. With humility you will build better networks, happier teams, and more valued relationships in your industry.

What Is Humility?

Humility is a much-misunderstood character strength. It is sometimes wrongly perceived as a weakness or a lack of self-worth. The Latin and Greek root of the word refers to "of the earth or ground" (*Humility*, 2022). It is defined as a balanced and accurate view of yourself, your accomplishments, and your identity. It is not self-criticism, self-effacement, or bowing to everyone else's demands. It is letting your accomplishments do most of the speaking about you and the ability to readily admit your own mistakes and shortcomings. A humble or modest person does not seek outward attention or praise and is more comfortable blending into the crowd. They also easily make friends as they are generally well-liked, helpful, agreeable, and generous. Humility is a prosocial character strength as it enhances social bonds.

The tricky part for those exhibiting great humility is finding the golden mean – the balance point between humbleness and a healthy need for recognition and appreciation. Music professions tend to striate along these lines. One might assume a performer lacks humility because of the nature of the "spotlight." To that end, one might assume a manager exhibits great humility by aiming the spotlight onto their artists. These are assumptions but are not diagnostic. Of course, a performer who exhibits low humility might be thought of as difficult to work with or someone who takes all the credit and doesn't share the limelight. The manager, whom neither seeks recognition nor shares the successes of their artists, will miss out on opportunities to grow their careers. It's all about finding and applying a healthy balance.

Humility and Your Brain

Humility is a highly desirable and sought-after character strength. There is more to this than the way you perceive yourself. It also extends to the way you think about others (Millacci, 2023). Studies have shown that gratitude and humility are mutually reinforcing. They have an upward spiral between them (Kruse et al., 2016). When you are aware of the ways others have helped you and are able to acknowledge that through words and acts of gratitude, you are growing your humility. Even the simple act of writing a letter of thanks helps improve your humility. Learn to see and speak the language of gratitude to grow your humility.

Amy Brann, author, and founder of the behavioral and neuroscience organization Synaptic Potential, lays out five hallmarks of humility (Brann, 2022). Humble people are secure and accept their identity. They are free from distortions, open to new information, other-focus, and have beliefs based on equality. She explains humility is an important core value to have in a company or organization because it facilitates positive, prosocial behaviors among employees and team members.

Affirmation

"I know who I am. I feel my own gifts and talents and I don't need recognition from outside sources to experience the meaning and joy they bring me."

Quick Tip

Embrace mistakes and imperfections openly as opportunities for growth and permission to be human.

Musical Examples of Humility

- Allowing others in the project to speak at the Grammy acceptance speech
- Acknowledging others' hard work in the program notes
- Letting the music speak for itself
- Disguising one's appearance or disappearing into the spectacle of the background in performance
- Thanking contributors for research in a book or journal
- Being in a band with no lead performer

Discussion Questions/Self-Reflections

Write about an area of your life where more humility would serve you well. How would you act, or what would you say or do differently? What would be the positive impact of this?

_____
_____
_____
_____
_____
_____

Treble and Bass

Treble • Overuse of Humility
SELF-DEPRECIATION: belittling your own accomplishment, not being able to acknowledge your own worth.

- Flutist who starts with an apology before playing the first note.
- Qualifying a performance by saying "I'm not much of a piano player."

Bass • Underuse of Humility
BASELESS SELF-ESTEEM: footless self-esteem; arrogance.

- Label executive who claims other people's success as their own.
- A mediocre artist who overvalues their ticket-selling draw in a venue.

Lights, Camera, Action!

- Let someone else take the solo.
- Support a deserving colleague's success today.
- Consider someone else's idea may be better than your own and go with it.
- Embrace and acknowledge a mistake you made – out loud.
- Give someone else a compliment instead of bragging about something you did.
- Ask someone you trust and who knows you to give you feedback on your struggles and areas of growing edges.

Humility VIP

Jeffrey Murdock • College Professor • Associate Director of Choral Activities

Humble people often don't call attention to their own achievements. They are doing what they do because it is of value to them, not for the glory of recognition. So it is outstanding when humble people are widely recognized for the greatness they have contributed to the world.

2021 Grammy Music Educator Award winner, Jeffrey Murdock epitomizes the strength and grace of doing ones job with humility (Liwag, 2021). As an associate professor of music education at the University of Arkansas, Murdock exhibited kindness and inclusion with his students, guiding them through the pandemic by bringing them together for a larger opportunity; to be part of an inspirational choir performance of "I Dream" by Florence Roach, honoring Dr. Martin Luther King (*U of A's inspirational chorale performs "I dream" to honor mlk*, 2021). Murdock spoke of his goals, "My passion is for leveling the field of music education and I'm passionate about increasing diversity, equity, and inclusion in the field." His service to underprivileged communities and his nurturing ways show humility in the way he teaches to uplift his students and bring out the best in them while keeping his own ego out of the way.

References

Brann, A. (2022, July 1). *Humility: The 5 hallmarks*. Synaptic Potential. Retrieved January 28, 2023, from https://synapticpotential.com/leadership/humility-five-hallmarks/

Kruse, E., Chancellor, J., Ruberton, P. M., & Lyubomirsky, S. (2016, January 19). *An upward spiral between gratitude and humility.* Social Psychological and Personality Science. Retrieved January 28, 2023, from https://escholarship.org/uc/item/77v6z086

Liwag, A. (n.d.). U of A's inspirational chorale performs "I dream" to honor mlk |. Retrieved January 28, 2023, from https://fulbrightreview.uark.edu/u-of-as-inspirational-chorale-performs-i-dream-to-honor-mlk/

Liwag, A. P. (2021, March 11). *U of a music's Jeffrey Murdock wins 2021 Grammy music educator award.* University of Arkansas News. Retrieved January 28, 2023, from https://news.uark.edu/articles/56221/u-of-a-music-s-jeffrey-murdock-wins-2021-grammy-music-educator-award

Millacci, T. S. (2023, January 6). *What is gratitude and why is it so important?* PositivePsychology.com. Retrieved January 28, 2023, from https://positivepsychology.com/gratitude-appreciation/

Wikimedia Foundation (2022, December 8). *Humility.* Wikipedia. Retrieved January 13, 2023, from https://en.wikipedia.org/wiki/Humility

Prudence

Musical Prelude to Prudence

Watch out for this one. It may seem less exciting and sexy than say, zest, but it's the Michelin star of strengths. In the same way a perfect meal is crafted with high-quality ingredients, they are nothing noteworthy without the expertise of a chef who knows how to deliciously balance flavor profiles and bring it all together with stunning presentation.

The discipline of prudence applied to art can make something everlasting. The best ingredients and intentions can pave the road to obscurity when a sound strategy is not in place. Creating something bold that moves the needle in the music industry takes patience and timing. So, if zest is the heat of creativity's fire, prudence is the exacting release of the flaming arrow.

We especially want prudence in particular music professions. For instance, we want our attorneys to be prudent and careful in our dealings. There are so many decisions to make in a music career, especially as an artist. Being an artist means you will face endless choices throughout your career – In which genre does my music best fit? Who should produce my album? Is this bass player up to par? Do I need a manager? How much should I charge for this show? The list goes on …

Some decisions bear heavier consequences and can affect your entire career. All too often, young or inexperienced artists are taken advantage of with terrible deals they signed because they were excited and eager to have the attention of a label, a talent show, or a manager. This is where prudence needs to unpack its bags and stay a while. One must carefully attend to the details, negotiate, and play out scenarios through time to see the real cost of opportunities. It's about choosing the right contract or job, not the flashiest, to best serve your long – and short-term needs and trajectories. And the career choices you make as a younger person will be different from the ones you lean towards as you age through your industry. You may find the full-scale jury of your mind is out for longer periods of time, carefully weighing evidence to make the best decisions as you progress.

And entities must also be prudent in their choices. Labels and publishers work with large sums of money that they are in essence investing in talent prospecting for future success. A label must be level-headed if they are spending their own time, money, and resources on something. It's way more than a question of talent or songs. If they don't have a prudent strategy to deploy, they can have a great artist, song, and product and it will sit on a shelf; everyone missing a big opportunity. The chef will have spent a lot of money on great organic ingredients that were left out too long and are no longer palatable.

So perhaps let's re-examine sexy. Is it not the prudent spy in the movie that lives until the ending credits? Even gets a sequel? Do we not admire those in the industry who are able to maneuver and morph through trends, technology changes, and decades to reach legendary status? Perhaps "play it cool, boy" is the wisest and sexiest advice we can receive.

What Is Prudence?

Prudence is one of the four temperance strengths that acts as a brake on the very human tendency of excess. Interestingly, most people aren't even sure what it is. People find it an odd, old-fashioned, unrelatable word. It is certainly not as exciting as the concepts of youthful exuberance and free-flowing expression. Prudence connotes non-sexy pragmaticism which feels flat and pales in comparison. Prudence gets short shrift and needs to be decontaminated to see its values (Thiele, 2006). Prudence is all about thinking and carefully weighing the consequences before speaking or acting. Prudence is also a large part of financial planning for future needs, rationing energy when running a marathon, going up for tenure, and allowing yourself the mental space and time to practice for that orchestral audition.

All change begins with prudently applying mindfulness and attention to a concern or problem. Prudence can be seen as a protector and filter that safeguards you against problems, dangers, and stresses in both physical and psychological areas of your life. It increases your level of conscientiousness and often leads to greater productivity. Along with optimism and sociability, prudent people exhibit less aggression and tend to have better physical health (Pappas, 2011).

Prudence and Your Brain

Prudence is most associated with personal internal qualities, and as such is mostly located in the areas of the brainstem and cerebellum.

The emerging behavioral neuroscience of prudence is very complex and is known to involve over a dozen brain regions and capacities (Arvan, 2020). What's going on in our brain when our behavior is cautious and prudent?

The prefrontal cortex, which makes us distinctly human, takes charge. It focuses our attention on thinking, planning, and reflecting, while it is monitoring and looking for errors. This is going on within a huge interdependent system. Nothing in our brain works alone (Calbet, 2022).

And prudence, as it turns out, is one of the "older" strengths, as our brains take time to fully mature to adult levels of self-regulation and more risk-averse behaviors. Just think about some of the decisions you may have made when you were younger and more "carefree" that you would not make right now (*The Science of Adolescent Risk-taking workshop report*, 2011).

Prudence is a key factor in having well-being, happiness, and making beneficial choices. It allows you to see further down the path to the long and short-term impacts of actions and decisions. It is also one of the self-regulators that is necessary in musical training. The beginning of the journey to learn any instrument requires a lot of faith that practice put in will result in mastery and even a career in music. Prudence helps you set your boundaries and clarify your choice making to guide you along the way.

Affirmation

"I am cautious, wise, and thoughtful in making my decisions."
"I breathe and take a mindful pause before speaking or moving into action."

Quick Tip

Watch, look, and listen first before responding – give yourself time and space to think and feel.

Musical Examples of Prudence

- Dedicating yourself to a sober or healthy lifestyle
- Hesitating to post something on social media
- Choosing criticisms carefully on an adjudication form
- Maintaining an emergency fund for your youth orchestra
- Choosing carefully which school to attend based on which has the best music business job placement
- Scheduling your studio sessions so that you do not burn out or run out of time
- Not taking an exciting opportunity to tour in favor of a more stable job offer
- Resisting over-valuing your new music tech business to a venture capitalist

Discussion Questions/Self-Reflections

What areas of your life could benefit from more prudence? How would this help you?

Which areas could use *less* prudence and what would be the impact of that for you?

_____
_____
_____
_____
_____
_____

Treble and Bass

Treble • Overuse of Prudence
STUFFINESS: rigid, closed mindset, easily offended when others talk about or do things you consider off-limits, private or crass, snobbiness.

- Middle school band teacher who can't laugh at a joke with their students.
- Music critic who takes an elitist tone and ignores the value of other music outside of their specialty.

Bass • Underuse of Prudence
SENSATION-SEEKING: risk-taking, throwing caution to the wind, thrill seeker, pathological spontaneity, "let's just do it" attitude.

- Musician who abuses drugs.
- Publicist who pushes outrageous stories about their artist to get attention.

Lights, Camera, Action!
- Take a mindful breath before sharing your opinion.
- Re-design a daily ritual that is not effective.
- Get in touch with your own artistic and professional values and thoughtfully choose artists for your label or publishing roster.
- Listen to all sides when friends or family members disagree.
- Reflect on a current situation or difficulty in your life and come up with five options on how to handle it.
- Plan an event or activity and spell out each and every step in detail. Think of possible obstacles and create alternate plan B's.
- Think of someone you know in your life that is prudent in ways you like. Describe these behaviors and their benefits.

Prudence VIP

Brian Epstein • Manager

One can almost describe a prudent person as having forward hindsight. They are considering all the angles and not making rash decisions, even when, at first glance, those seem to be the shinier choices. Brian Epstein's vision and prudence is credited with bringing the Beatles to the world stage (Lewis, 2017). As their manager, he first spotted the local act and saw their potential and charm. Though he had no experience with management, he doggedly worked to secure for them both opportunities and contracts. His careful advice influenced the Beatles in all manners – style, dress, presentation, and public appearances; however, he wisely stayed out of their musicianship. This very prudent choice built mutual trust and respect between the two entities. Brian Epstein went on to manage the Beatles, guiding them through their artistic evolution and supporting their journey, until his death in 1967.

References

Arvan, M. (2020). *Neurofunctional Prudence and morality: A philosophical theory*. Routledge, Taylor and Francis Group.

Calbet, J. (2022, May 6). *Prudence: Strengths and opportunities for improvement*. Neuroquotient. Retrieved January 28, 2023, from https://neuroquotient.com/en/prudence-strenghts-and-oportunities-for-improvement/

Lewis, M. (2017, December 7). *No Brian Epstein? no beatles!* HuffPost. Retrieved January 28, 2023, from https://www.huffpost.com/entry/no-brian-epstein-no-beatl_b_5155998/amp

National Academies Press (2011). Institute of Medicine (US) and National Research Council (US) Committee on the Science of Adolescence. The Science of Adolescent Risk-Taking: Workshop Report. Washington (DC): National Academies Press (US). Available from: https://www.ncbi.nlm.nih.gov/books/NBK53418/ doi: 10.17226/12961

Pappas, S. (2011, March 15). *Hard-working and prudent? You'll live longer*. LiveScience. Retrieved January 13, 2023, from https://www.livescience.com/13258-hard-workers-live-longer.html

Thiele, L. (2006). *The heart of judgment: Practical wisdom, neuroscience, and narrative*. Cambridge: Cambridge University Press. doi: 10.1017/CBO9780511498718

Forgiveness

Musical Prelude to Forgiveness

How heavy is the load you carry? It often feels in music professions you live many different lives, sometimes simultaneously. You may start in one field, and gravitate towards another, all the while trying your hand at something on the side. These different terrains bear entire networks of different contacts. It's like being a resident of multiple neighborhoods at once. In all those dealings, you are bound to have both positive and negative transactions and you will learn where and when not to tread. You will develop trusted connections that you return to repeatedly throughout your growth and people will see you in a like manner. It would be foolish to assume nothing will ever fall through, go gloriously wrong, or end with bad blood.

Sometimes decisions are made, and contracts are signed, in the early stages of someone's career that aren't ideal and don't grow or support the artist with time.

They weren't borne of malice but rather an inexperience, hastiness, or simply not having any upper hand in the matter. The feeling of a past decision like that can change as an artist develops and moves into higher echelons. It's easy to hold that against someone and live with regret or buried anger. Forgiveness allows you to reframe old feelings and release a bevy of energy that can be spent on building something new or better. Sometimes you must let things go to move forward and you may not know the burden and real cost of holding onto a grudge. Forgiveness is about recalibrating our energies and learning where they are seeping out. We can leak a lot of our energy if we are carrying contracts of old bitterness and not even know it. We may just feel depleted.

It's important to see things through the other person's eyes, circumstances, values, and strengths. Practicing forgiveness is an admission that you too have faults. If given the reverse situation, you would want someone else to pardon your misjudgments. It's an act of humility and self-regulation to find it in your heart to see the other side. Opening yourself empathically to others grows your tolerance and understanding. As you develop blisters on your feet from walking in the many shoes and bare feet of humanity, you will see and feel things from all sides.

People also have the capacity to grow. The villain of your memories who wronged you may have undergone a whole re-evaluation of their own behaviors. We don't know the other person's thoughts or suffering. Or even perhaps worse, we are the only ones suffering someone's long-ago slight and they don't even recall the violation. That means that the pain has been constantly just affecting you – so you endure it over and over.

An important component of forgiveness is honesty. First, to acknowledge precisely how something has affected you and why it bothers you. You may find it rubs up against one or more of your signature strengths. Next, you need to find a version of peace by letting the other person know how they have violated you or crossed your boundary. This is not always possible or prudent. There are techniques for writing and burning letters without sending them that allow you to release bottled-up emotions and move on. It might feel like draining infection out of a wound to forgive. And that is the point – your human energy is the most precious commodity you have – how do you want to spend it in your lifetime?

What Is Forgiveness?

Forgiveness is a powerful healing strength, as much for the giver doing the forgiveness, and the person being forgiven. Forgiveness is a choice to change your mental state and decide to release anger and bad feelings toward another person who deliberately or unintentionally did something unfair or harmful to you. Forgiveness holds hands with mercy – and involves accepting the flaws and imperfections of others and yourself as you give them another chance.

Forgiveness does not mean that you forget or condone bad behavior. Forgiveness does not require that you stay or re-engage in the situation or in the relationship. It is an act of self-compassion and self-care that greatly relieves the carrying of old burdens or ancient bad feelings. Forgiveness is more of a psychological internal response whereby you let go and rise above the negative judgments and resentments toward the individual that hurt you (Souders, 2022). You see the situation from a different perspective – more benevolently – acknowledging the suffering and struggles of all of mankind, even the offender. You chose to feel compassion for both you and the other.

It is challenging to find forgiveness for someone who has hurt us. It involves extending understanding and empathy. Psychologist Shauna Shapiro writes, "Forgiveness is perhaps the most challenging of all the resources available to us–and the most transformational" (Sutton, 2020). This is because forgiveness chooses to let go and choose mercy over vengeance. Think about how difficult forgiveness plays out in the creative fields. There are bands that hold grudges years beyond the conflict that broke them up. Other bands have reunited and chosen to "do the work" of forgiveness.

Psychologists maintain that harboring a grudge is not beneficial to one's emotional or physical health (Vanbuskirk, 2021). Stanford psychologist and forgiveness expert Fred Luskin reported that the mental benefits of forgiveness training are increased optimism, self-confidence, compassion, reduced stress, and spiritual inclinations. And these benefits were still present six months after training (Steakley et al., 2017).

Many people confuse forgiveness and spiritual bypass (Blythe, 2022). "Spiritual bypass" refers to cutting off your feelings and believing prayer or God will fix everything. Forgiveness does not attempt to ignore suffering but instead strengthens your capacity to remove emotional barriers to finding happiness. Buddhist psychology sees forgiveness as a powerful character strength that provides a direct way to reduce our suffering and bring us dignity and harmony in our life (Ricci, 2019).

Dr. Tal Ben-Shahar uses the phrase "Permission to be human" (*Tal Ben-Shahar on how to be happy*, n.d.). This is a mantra in positive psychology that gives people the grace and permission to be flawed and imperfect. Let this become a personal mantra and when you make a mistake – use it to move into forgiveness. This practice connects you in a human way with all of mankind and counteracts our tendency to feel alone and flawed.

Forgiveness and Your Brain

Granting forgiveness is associated with activating a brain network involved in empathy and the regulation of affect through cognition. On a structural level, this involves the precuneus, right inferior parietal regions, and the dorsolateral prefrontal cortex (Ricciardi et al., 2013).

In studying forgiveness, researchers have found that the tendency to forgive is linked with certain structural and metabolic characteristics of the brain (Moawad, 2020).

The structure and size of different parts of the brain are tied to differing levels of forgiveness. One can be born more forgiveness-prone than another. However, studies on the differences in metabolic brain activity associated with forgiveness suggest that this strength can change throughout life. As with all character strengths, it can be taught and cultivated through practice (Li & Lu, 2017).

Forgiveness provides vital benefits to your mental and physical wellbeing: a reduction in depression, anger, stress, pain, and cardiovascular disease. As well, forgiveness increases and improves hope, compassion, self-confidence, and your body's immune responses. There is no downside to forgiveness (*Forgiveness: Your health depends on it*, 2021).

Affirmation

"I forgive others when they behave badly or upset me. I process that information to guide me in my future relations with them."
"I forgive myself when I make mistakes, face my limitations, or hurt someone else. I combine my forgiveness with self-compassion."

Quick Tip

Just let it go!

Musical Examples of Forgiveness

- Calling a band member after a fight
- Allowing for the humanity of a bad moment in a rehearsal
- Providing training to a teacher who made a rookie mistake
- Deciding to manage a notoriously difficult artist after their rehab
- Producing an ex's music
- Setting up healthy boundaries with a toxic band member
- Getting the band back together after a lengthy breakup
- Renegotiating a contract with a label with whom you were at odds

Discussion Questions/Self-Reflections

Forgiveness is really about you – letting go of a burden or taking back your power. It releases the negative energy and attachment you have with the situation or the person. Write about a situation in your musical life that could shift and benefit from forgiveness. How do you think finding forgiveness in this situation would help you in the larger picture? What would be the benefits to you and others involved?

_____
_____
_____
_____
_____
_____

Treble and Bass

Treble • Overuse of Forgiveness

PERMISIVENESS: an extreme allowance of things that others might not allow, perhaps letting others step past your own boundaries, not standing up for yourself.

- Junior high school choir director who allows the classroom to become unruly.
- A performer who accepts a lesser than agreed-upon fee after a gig.

Bass • Underuse of Forgiveness

MERCILESSNESS: unforgiving of others' mistakes, showing no pardon, pitiless.

- Church musician who holds a grudge because a member of the band switched churches.
- Music director who berates musicians in front of a crowd because they are playing the wrong chords.

Lights, Camera, Action!

- Get in touch with someone from your artistic past where it didn't end well and send a sincere message letting go of old bad blood.
- Forgive yourself for anything you felt was not good in the past (a substandard performance, a bad business choice, a low-quality recording). Move forward, honoring the lessons of the past.
- Write a letter of forgiveness to someone. Send it or burn it.
- Write a letter of forgiveness to yourself. Read it aloud and really take it in.
- Engage in a joyful activity you have blocked because of fear or shame over a past failure. Could forgiveness help you find more joy?

Forgiveness VIP

Selena Gomez
Artist • Singer • Actress

Holding onto anger, even when justified, can feel toxic. Forgiveness is a grand act of accepting that humans are flawed and will make mistakes and have the capacity to hurt us. But we hurt ourselves a second time by holding onto the negativity that ensues from an offense. From her public relationships and breakups to health issues to her enormous social fame and living her life in the spotlight, Selena Gomez has spoken publicly on the importance of forgiveness (*Selena Gomez: It's best for me to turn the other cheek*, 2015). Speaking about the lyrics of her song "Kill Em With Kindness," Gomez explained that people who say hateful things about her online don't deserve the energy of retaliation. She says the best thing for her is "turning the other cheek." Eventually, she realized it was mentally healthier for her to distance herself from her phone and all her followers to gain inner mental peace. Selena practices forgiveness along with acceptance and self-kindness. When she learned of her bipolar disorder, she felt relief that she finally understood herself better. Selena's resilience through multiple ailments including lupus, a kidney transplant, chemotherapy, and mental illness speaks to her inner strengths pulling her through (Brockington, 2021).

References

Blythe, A. (2022, October 27). *Spiritual bypass harms victims.* Betrayal Trauma Recovery. Retrieved January 28, 2023, from https://www.btr.org/spiritual-bypass-harms-victims/#:~:text=Sometimes%20these%20emotions%20become%20so,to%20experience%20and%20process%20them

Brockington, A. (2021, August 19). *Selena Gomez opens up about health struggles in the public eye: "The narrative was so nasty".* The Hollywood Reporter. Retrieved January 28, 2023, from https://www.hollywoodreporter.com/lifestyle/lifestyle-news/selena-gomez-health-struggles-1235000041/

Forgiveness: Your Health Depends on It | Johns Hopkins Medicine. (2021, November 1). *Forgiveness: Your health depends on it.* Retrieved January 28, 2023, from https://www.hopkinsmedicine.org/health/wellness-and-prevention/forgiveness-your-health-depends-on-it

Li, H., & Lu, J. (2017, November 20). *The neural association between tendency to forgive and spontaneous brain activity in healthy young adults.* Frontiers in Human Neuroscience. Retrieved January 28, 2023, from https://pubmed.ncbi.nlm.nih.gov/29209186/

Moawad, H. (2020, November 14). *The neurobiology of forgiveness.* Neurology live. Retrieved January 28, 2023, from https://www.neurologylive.com/view/neurobiology-forgiveness

Ricci, J. (2019, July 11). *The practice of forgiveness.* Jack Kornfield. Retrieved January 28, 2023, from https://jackkornfield.com/the-practice-of-forgiveness/

Ricciardi, E., Rota, G., Sani, L., Gentili, C., Gaglianese, A., Guazzelli, M., & Pietrini, P. (2013, December 9). *How the brain heals emotional wounds: The functional neuro-anatomy of forgiveness.* Frontiers in Human Neuroscience. Retrieved January 28, 2023, from https://www.ncbi.nlm.nih.gov/pmc/articles/PMC3856773/#:~:text=Granting%20forgiveness%20was%20associated%20with,and%20the%20dorso-lateral%20prefrontal%20cortex

Music (2015, September 26). *Selena Gomez: It's best for me to turn the other cheek.* Retrieved January 28, 2023, from https://www.music-news.com/news/UK/92480/Read

Souders, B. (2022, July 25). *What is forgiveness? (+9 science-based benefits).* Positive-Psychology.com. Retrieved January 19, 2023, from https://positivepsychology.com/forgiveness-benefits/

Steakley, A. L., D'Ardenne, A. K., & Armitage, A. H. (2017, December 19). *Stanford psychologist Fred Luskin taking questions on the health benefits of forgiving.* Scope. Retrieved January 28, 2023, from https://scopeblog.stanford.edu/2012/01/23/stanford-psychologist-fred-luskin-taking-questions-on-the-health-benefits-of-forgiving/

Sutton, J. (2020, September 3). *Psychology of forgiveness: 10+ fascinating research findings.* PositivePsychology.com. Retrieved January 28, 2023, from https://positivepsychology.com/psychology-of-forgiveness/

Octavian Report (n.d.). *Tal Ben-Shahar on how to be happy.* Retrieved January 19, 2023, from https://octavianreport.com/article/tal-ben-shahar-on-how-to-be-happy/

Vanbuskirk, S. (2021, August 19). *The mental health effects of holding a grudge.* Verywell Mind. Retrieved January 28, 2023, from https://www.verywellmind.com/the-mental-health-effects-of-holding-a-grudge-5176186

THE WISDOM STRENGTHS

Curiosity

Musical Prelude to Curiosity

Remember when most of the world was still a mystery? When there was a wide world of wonder on the other side of every door? Recall the first time you were completely drawn into a band's music as if you were overwhelmed by their interesting sonic kaleidoscope. Or the first day of an internship at a record label as you listened to the rapid and strange lingo being exchanged which ignited specialists to dive into work tasks. Or your first foray into playing with the capabilities of your new synthesizer. There was more stretch of the shadowy, enticing forest ahead of you than behind you and you were craning your neck to see as much as you could through the leaves.

The newness of the unknown or unexplored is the siren song of curiosity. It is the call that brings us deep into the nooks and crannies of our industry and serves as the mother of invention. To be curious is to almost *need* to know more about a given subject. First, you fell in love with the band, then you learned how to play their music and now they inspire your production ideas. In the first label job you had, one of the specialists took you under their wing and today, you head up that division. And that early synthesizer became a means for you to tell your sonic story, which in turn got you into film scoring. Curious people take that spark of interest and walk it down the path, illuminating other enticing trails begging to be explored.

In all facets of the music industry – whether creative, technological, business, or medical – it is so important to be curious about what's coming next. It is the way to stay relevant but more importantly, to do no harm. It serves no one to sell stale bread and those who refuse to pay attention to the changing times can actually hurt people with outdated facts and practices. Students pay a lot of money to learn how to develop in the industry of tomorrow, not even today and certainly not yesterday. Teachers need to be curious enough to keep current so that they can deliver the freshest product. Money talks in the music industry, and those who refuse to stay curious and up to date will pay with real currency. And, of course, music therapists need to be active in the development of their field so that they can apply new findings to patient care.

Curiosity is a companion of creativity. The many fusions and interesting developments through time in composition, songwriting, and production came from challenging the standard and asking, "What would it sound like to layer this influence over this one?" There were vanguards who put it all on the line to throw something original out into the culture. They knew the response would be binary in nature – people would love it or hate it. And that was a worthy risk; a winner takes all. Think of the many artists who challenged the stereotypes, pushed the line, and came out of the gate swinging with their own creative, curious self-expression. What comes after that kind of novelty is a lot of copycats who are inspired or simply want to replicate success. And there will always be artists who prefer not to look back and label the past their "glory days." Instead, they choose to lean forward, peeking through the shadows of the forest, asking the bravest question – what's next?

What Is Curiosity?

The VIA describes curiosity as being open to different experiences and activities and taking an interest in them for their own sake. Curiosity is the "tell me more" strength. It is a combination of hunger for novelty, persistence, and intelligence (*Intelligence*, n.d.).

The origin and development of mankind is steeped in curiosity. Just imagine the thrill, fear, and delight the first-time humans discovered how to create fire. Think of all the subsequent inventions and developments that followed all borne out of curiosity (*Encyclopædia Britannica The Science of Curiosity*, n.d.). This special spark of curiosity drove humans to explore, discover, and invent.

Curiosity is the essence of human existence. It drives the questions, "Who are we?," "Where do we come from?," and "How can we find meaning in our lives?"

Socrates said, "The unexamined life is not worth living" (*The unexamined life is not worth living*, 2022). This is more of a reference to curiosity than knowledge. In every society, there are curious members on the leading-edge pushing knowledge forward fueled by curiosity.

You could say that curiosity is the main driver for human flourishing. Some have called curiosity the engine of achievement. Stephen Hawking, well known for his brilliance and creativity in quantum physics, reminded us to stay curious in an interview with Diane Sawyer, "Remember to look up at the stars and not down at your feet. Try to make sense of what you see and wonder about what makes the universe exist. Be curious" (*ABC World News with Diane Sawyer: Conversation with Stephen Hawking*, 2010).

We are deeply curious beings. Our lives and society are shaped so strongly by this drive to obtain information that we are sometimes called informavores – creatures that seek and digest information, just like carnivores eat and hunt meat (Machlup & Mansfield, 1984).

Curiosity connects us to others in a joint pursuit of knowledge. It is a key part of learning at all ages. The minds of curious people are always active, wanting to know and understand their world. They also tend to be more open and eager to learn from others.

However, curiosity can be overused when it shows up as prying into another person's life. Think of how abusing curiosity at work or in a band can lead to prying and gossip. It's good to balance your curiosity with your prudence and self-regulation.

Curiosity and Your Brain

Exactly how does curiosity work in our brains? When something piques your interest, your brain enters the "curiosity state" (*How curiosity changes the brain to enhance learning,* 2014). Parts of the brain sensitive to unpleasant conditions light up signaling that you are lacking certain knowledge. This triggers the parts of your brain responsible for learning and memory to kick into high gear. At this point, you are optimally primed to begin searching for answers. And when you begin learning new facts in this "curiosity state," something even more interesting happens: your reward circuitry kicks in. Your brain floods you with dopamine to reward you for being curious and for pursuing it! At the end of this cycle, you are left feeling happier. Just like we train our pets with rewards and treats, your brain remembers this pleasantry and you will be incentivized to be curious again.

To understand this more, we must understand what it means to be "information seeking." All animals seek information about their surroundings as a means of safety and self-preservation. The senses are invaluable guides through this process, constantly feeding information back to our brains (*Encyclopædia Britannica The Science of Curiosity,* n.d.). Trusting and relying on your sensory abilities – seeing, hearing, smelling, tasting, and touching – are the main ways we gather and interpret the data of what we are curious about.

While in modern society, our curiosity isn't as vital to our survival, it is useful for our education, happiness, and well-being. Being curious can hugely enrich your life and guide you to pursue passions for both short- and long-term satisfaction.

Most geniuses and intellectual giants – Albert Einstein, Marie Curie, Neil deGrasse Tyson, and Leonardo DaVinci – have the trait of curiosity. Curiosity is a mental exercise that literally strengthens your mind, keeping it active over time. It makes your mind more observant of new ideas and they become easier to recognize, opening new worlds of possibilities that are often hidden below the surface of normal life. Having a curious mind brings color and excitement into your life, finding wonder in the everyday.

You can explore your curiosity through the study of topics that interest you and through social involvement. And there is nothing that stimulates our curiosity more than a complex topic shrouded in unknowns.

Affirmation

"I follow my curiosity and ask lots of questions."

Quick Tip

Questions are your best tool for connection and engagement. Curiosity = Happiness

Musical Examples of Curiosity

- Researching biographical aspects of musicians or music history
- Exploring music outside of your worldview
- Pursuing a deeper understanding of music theory
- Learning how to play other instruments
- Exploring foreign sounds or traditions
- Connecting to fellow musicians who you find interesting
- Experimenting with different or extended techniques
- Analyzing musical scores
- Testing new formal or sonic theories
- Building/designing music tech
- Producing new sounds or combinations
- Musicological research

Discussion Questions/Self-Reflections

Would you rather be asked questions or be the one who is asking? Try on both roles and notice what each is like for you and how they are different. Which do you prefer? Decide to ask two questions you are curious about with at least three different people and write about that experience.

_____
_____
_____
_____
_____
_____

Treble and Bass

Treble • Overuse of Curiosity
NOSINESS: you ask so many questions hardly pausing to acknowledge reception. The questions seem to be more about you than getting to know the other person.

- Music attorney who breaches client privacy by gossiping.
- Opera director who delves far into the personal lives of the cast.

Bass • Underuse of Curiosity
DISINTEREST: the only reason people do not know much is that they do not care to know. They are uncurious.

- Label president who does not keep up with the latest trends.
- French horn player who doesn't want to spend time learning about proper posture for playing.

Lights, Camera, Action!

- Go to a restaurant that plays the regional music of the food being served and immerse yourself in the experience.

- Ask your mentor questions you have always wanted to know about them.
- Read a book on a subject that interests you.
- Get up from behind your computer and explore a physical bookstore or library and browse the shelves. Google searches find specific information related to questions, but we are unlikely to have serendipitous encounters in natural ways fueled by our own senses and our curiosity.
- Watch a mystery movie and try to guess "who did it"!

Curiosity VIP
Danny Elfman • Film Scorer

It is said that curiosity killed the cat. For those of you who have cats, you've probably witnessed that one-step-too-far your furry friend endeavored, leading them to a teetering place or a dangerous jaunt. Curiosity entices the mind to go down hallways of beckoning unopened doors. Composers who deal with music and visuals are encountered with extra-musical cues and oftentimes must prepare by researching unknown subjects, studying characters, or perhaps visiting new places (Romain, 2019). Film scorer Danny Elfman is famous for creating lush and mystical musical landscapes and applying magical sounds to fantastical stories. His curiosity not only shows in his ability to compose novel sounds for fictional worlds but also in his lifetime of traveling to exotic places and collecting interesting items. He even speaks of creating stories around his collected "curiosities," exemplifying another strength – creativity.

References

Encyclopædia Britannica, Inc. (n.d.). *The Science of Curiosity*. Encyclopædia Britannica. Retrieved January 28, 2023, from https://curiosity.britannica.com/science-of-curiosity.html

Machlup, F., & Mansfield, U. (1984). *The study of information: Interdisciplinary messages.* Wiley.

Romain, L. (2019, September 12). *Danny Elfman takes us through his collection of oddities*. Nerdist. Retrieved January 28, 2023, from https://nerdist.com/article/danny-elfman-creepy-collection-instagram/

ScienceDaily (2014, October 2). *How curiosity changes the brain to enhance learning*. ScienceDaily. Retrieved January 28, 2023, from https://www.sciencedaily.com/releases/2014/10/141002123631.htm

Sussex Publishers (n.d.). *Intelligence*. Psychology Today. Retrieved January 28, 2023, from https://www.psychologytoday.com/us/basics/intelligence

Wikimedia Foundation (2022, December 23). *The unexamined life is not worth living*. Wikipedia. Retrieved January 28, 2023, from https://en.wikipedia.org/wiki/The_unexamined_life_is_not_worth_living

YouTube (2010). *Abc world news with Diane Sawyer: Conversation with Stephen Hawking*. YouTube. Retrieved January 28, 2023, from https://www.youtube.com/watch?v=MJBwKCHjlXI

Creativity

Musical Prelude to Creativity

Creativity is a fire that burns within all of us. As life happens, that flame risks being burned out by the experiences and cycles we go through. Naturally creative people have innovation always running in the background, like an app you forget to close out. They filter life through a different lens on purpose. They are constantly looking for clever connections between seemingly disparate events or visuals. Since it's at the heart of music making, one might assume all musicians feel very creative. This is not so. There are many ways in which people participate in music. Many musicians are not trained to put themselves into the music a.k.a. create. They instead are trained like finely tuned machines that can execute incredibly difficult passages under the pressure of thousands of eyes and ears. In the same way, Olympians might feel more like machines than creators. Musicians commonly come to crossroads and find themselves questioning their place in the music. Maybe they are not given enough space to interpret, or they have detailed constraints on their musical flow due to assignments, deadlines, and extra-musical instructions as in film scoring, playing the music of long-gone composers, and writing for artists. Perhaps they have undergone a significant life circumstance change and the music they were making or playing no longer reflects their new reality. Whatever the cause, musicians frequently have a creative crisis and must reconnect to their core to find their way through. They must reignite their fire and find new ways to fan the flame.

Creativity is showing your soul respect. It's the assertive, inner voice demanding to be heard on the outside. Self-expression, assembling ideas in a novel way, and asking crazy questions of "what if's" are all part of strengthening the creative impulse and learning to trust it. In an effort to conform or be polite, creativity can sometimes be too quiet. It can get stifled by the noise of rules, norms, "shoulds," and "musts." It might look nothing like what you are told to do to reach success and may even forcibly buck the conventions altogether. Despite these risky maneuvers to get to creativity, people love to celebrate it when it wins big. We uphold those who have gone against the grain but sometimes only when they are accepting an award or gaining status. The trophies and regalia are hard-won relics, and the full story often goes untold. Winning a creative battle takes real courage, perseverance, and zest. It takes the bravery to champion something that doesn't yet exist – and it may be you alone and your own solitary idea.

Creativity in the music business is a replenishing task. One must continually find new and novel ways to get ears and eyes on art. You need to find creative avenues and angles to retell the story of another artist or album release in a slightly different way each time. But truly creative people love this challenge! Give one of these people limitations and their eyes widen with excitement. The limitations will force out the obvious answers and force true imaginativeness.

Regaining creative control of your own musical output and decisions are the goals of many artists who have signed them over to other entities. Even owning your own catalog is a symbolic and real power that aligns with claiming one's own creative voice.

So why do we place such a high value on creativity? Perhaps the glowing light inside the sacredness of creativity is the fact that we are all singular and unrepeatable. LIke the uniqueness of your VIA CS profile, no one will ever see the world the way we particularly see it and no one can build the unique fire that comes from a soul's yearning. The creativity and music we leave behind are our holy relics.

What Is Creativity?

According to the VIA, creativity is thinking of new and better ways of doing things. A creative person is not content with robotically doing things in the same familiar and conventional ways. Creativity is the power of your imagination.

Remember when you were young, and you just used crayons and markers however you wanted? The final product mattered less than the delight of playing with the colors and materials. You were engaged in an experiment of creative play. The dilemma is that often the older and more trained we become, the easy flow and spontaneity of our youth diminish. To recapture that kind of unselfconscious creativity, you've got to get out of your head and just do it for the sake of the here-and-now experience. Don't even think of a goal, project, or endpoint – discover your gifts by trying new things and experimenting. Go wide and expansive – just say "Yes, let's try that on!" Let your creativity be messy and driven by your imagination, your senses, your dreams. Play with what you don't know or do routinely.

When you dare to see things in novel ways, you use your unique creativity to bring something different into the world that is distinctly your own!

Creativity is not the story of the lone genius; it is the result of connectedness (Proudfoot & Fath, 2021). When you talk to creative people and ask them, "How did you think of that?," often, with some guiltiness, they say they didn't really do anything; they just saw or heard something with new eyes or ears. Eventually, connections became obvious, and they were able to synthesize previous experiences into new things. This kind of automaticity happens a lot with songwriting and composition. These creatives will often speak of being "used" to "channel" something or the music using them as a conduit or medium. Performers speak of these phenomena as well – feeling as if the music is playing through them, rather than playing it themselves. Some of these are described as spiritual experiences. Also, being part of an artistic community has an osmosis factor that invites expansive creativity and an open-growth mindset.

Creativity and Your Brain

English psychologist and neuroscientist Peter Malinowski says, "To maximize your brain's creativity, your job is to collect good ideas. The more good ideas you collect, the more you have to choose from and to be influenced by" (Christensen, 2017). Think of the way jazz musicians build their improvisation bank by listening to and playing with great musicians. Or the way children learn how to put different inflections into their singing voices by listening to artists they love and emulating them. And studying music in college is a laboratory of this creative collection and social engagement.

Pay attention to your choices: where do you direct your attention? How do you take in and capture details of what you see, hear, and experience? Being mindful of the richness of life's details can help you grow your creativity. Harvard educator and positive psychology author Dr. Tal Ben-Shahar teaches that "at every moment you have a choice" (Ben-Shahar, 2012). How could you apply this positive psychology concept to recharge your creativity?

You can combine all this playfulness with discipline. If you sit down in the same place every day and just make something happen, something will emerge. It's what Michelangelo meant when he said that the sculpture was in the stone; it was his job to release it!

Creativity thrives in wild open spaces – let your creativity roam free and breathe! We often too quickly push through deadlines to finish things and get them off our plates. Creativity is a strength that needs space, some contemplation, lots of air, and human interconnections to thrive.

Affirmation

"I am creative, and I have my own special gifts. I am open to inspiration, and I am happy to be myself."

Quick Tip

Just do things differently. Be a kid and shake it up just for fun. No thinking allowed – just play!

Musical Examples of Creativity

- Writing your own song
- Dressing up for performances
- Staging yourself creatively
- Thinking of new ways to perform traditional music
- Involving unexpected elements
- Responding with excitement to limitations
- Finding novel combinations of sounds
- Producing your own music
- Using instruments in unexpected ways
- Finding interesting places to perform
- Using unusual combinations or fusions
- Exploring musicological topics from unexplored perspectives

- Giving out-of-the-box assignments
- Finding novel ways to assess students

Discussion Questions/Self-Reflections

Think of the most creative individual you know and describe their characteristics. Adapt one of these behaviors and experiment with it. Think of a time when you used creativity to accomplish something important. Write down two specific ways you used this strength and what it felt like.

_____
_____
_____
_____
_____
_____

Treble and Bass

Treble • Overuse of Creativity

ECCENTRICITY: your creativity never stops – it exhausts others, and you feel like no one keeps up with your constant flow of the new.

- Going too far from the instructions for a film score.
- Inability to write within a song form or follow instructions.

Bass • Underuse of Creativity

CONFORMITY: just doing things in the same way – every time – like Groundhog's Day.

- Unable to bring forth uniqueness to a vocal performance.
- Unwilling to break from the mold of a school organizational structure that is outdated.

Lights, Camera, Action!

- Infuse a theme into your next show and pull it through the whole experience.
- Take items out of your pockets or purse and write a song connecting them into a larger narrative.
- Design/create/choose a culture of people for yourself that support your creativity.
- Embrace new ideas with an open mindset so that they can blossom.
- Take risks – experiment and fail more!
- Think/make several iterations of the same thing – turn it upside down.
- Look for exemplars and read or watch biographies about their lives – notice how they think and approach life and deal with obstacles and conflict.
- Pay attention to how you talk to yourself – be kind and easy with yourself.
- Think thoughts and feel feelings that combine and serve you in a positive way.

Creativity VIP
David Bowie • Artist

The most iconic and enduring artists in a lifetime have the rare ability to be fresh, honest, innovative, and adaptable to the times. Those lucky enough to lead the pack are often examples of extremely bold and creative people who are not afraid to be themselves or take risks with their music, their fashion, and their performances. David Bowie (*David Bowie*, 2023) is an icon who personifies the best of creativity. Just Google Ziggy Stardust (Bertram, 2020), a fictional character Bowie created as an alter-ego, gender-fluid, alien rock star, and you can begin to see why he was so revolutionarily creative. Even Bowie's methods of brainstorming were creative. He was known to practice a "cut-up technique" whereby he tore printed type into pieces and rearranged them in an order to create a new meaning (Chicola, 2020). He loved the challenge of associating often totally disparate ideas from this technique and it led him to bizarre and interesting concepts and lyrics. His creativity is his legacy as he was not only an ingenious performer, songwriter, and musician but also an actor and painter.

References

Ben-Shahar, T. (2012). *Choose the life you want: 101 ways to create your own road to happiness*. Experiment.

Bertram, C. (2020, December 17). *How David Bowie created Ziggy Stardust*. Biography.com. Retrieved January 29, 2023, from https://www.biography.com/news/david-bowie-ziggy-stardust

Chicola, J. (2020, February 19). *How a version of David Bowie's brainstorming technique can boost your… …* Retrieved January 29, 2023, from https://www.inc.com/jason-chicola/david-bowie-cut-up-technique-creativity-boost.html

Christensen, T. (2017, November 28). *Most everything you need to know about creativity*. Creative Something. Retrieved January 29, 2023, from https://creativesomething.net/post/130616843175/most-everything-you-need-to-know-about-creativity

Proudfoot, D., & Fath, S. (2021.). *Personality and Social Psychology Bulletin* Retrieved January 19, 2023, from https://journals.sagepub.com/doi/10.1177/0146167220936061

Proudfoot, Devon, & Fath, Sean (2020). Signaling creative genius: How perceived social connectedness influences judgments of creative potential. *Personality and Social Psychology Bulletin, 47*, 580–592. 10.1177/0146167220936061

Wikimedia Foundation (2023, January 24). *David Bowie*. Wikipedia. Retrieved January 29, 2023, from https://en.wikipedia.org/wiki/David_Bowie

Love of Learning

Musical Prelude to Love of Learning

Life is change, growth, and the accumulation and application of reems of knowledge. Though there are instruments that one can pick up and immediately use, like the voice, the possible depth of musical learning is endless. One can sit anywhere on the spectrum from bedroom guitarist to learning the most intense riffs and scales and extended harmonies. One can also widen the yoke to include the historical aspects of the instrument, fretboard and keyboard harmony, scoring, notation, chart writing, sight-reading, and endless books of repertoire from different styles and time periods. The love of learning is a spoonful of sugar helping all this medicine feel, well, sweeter. Think of this strength as skis to smoothly travel down the mountain of knowledge and make the steep learning curve feel exciting rather than intimidating. This strength holds hands with curiosity but puts it on a more rigorous and structured path.

People who truly love to learn get excited about new challenges and information. They are endlessly hungry for self-improvement and edification in this way. Details matter and make for interesting conversations. Lifelong learners also tend to teach and give back as much as they take in. Whether it is formal or informal, they are the people who have deep-cut information on areas that interest them. Talking to a person who knows droves of facts about the personnel, details, and events surrounding a band or musical period feels like stepping into a well-done documentary. They can deliver knowledge, oftentimes in a storytelling way, that belies their true enthusiasm and love!

Lovers of learning commit to the process and respect the steps it takes to have mastery over a subject. This may be evidenced by looking at their walls. These scholars and researchers tend to collect the academic swag of degrees, certificates, conference badges, and a dossier of publications with their names on them. They live in the present and are not scared of change. Instead, they are excited by it – it presents more opportunities to learn. They have a warm feeling about things they don't know and get psyched about the growth potential of new jobs, positions, roles, and tasks. They love a challenge in this way and are proud of themselves when they do master something new. And then it's on to the next.

You may notice that some musicians are multi-instrumentalists who dive into learning new instruments across a wide spectrum and variety. They also are not shy about asking for help, taking lessons, and frankly being bad before they can be good. There is humility in those who are willing to be joyful beginners at something even when they are quite expert at other things. They also champion and protect other learners and might be the guiding voice you seek when struggling with changing careers, especially when that move would require more education.

In business, those who don't love to learn are left behind merely holding the knowledge they ingested in their training or education only. They don't update their operating systems. By not voraciously dining on news and facts, they are incapable of presenting a balanced and modern opinion or perspective.

So, fall in love with learning. It's a strength that will joyously propel you down your long and winding path through endless lifetime expansion, no matter the musical field. Boosting your love of learning can only make the journey more fun and enriching and sweeten the many hours, tasks, and lessons you must master to get to your personal next level.

What Is Love of Learning?

If love of learning is a top strength of yours, it may have even driven you to read this book! You love learning new things, whether formally or just following your curiosity to explore your world. You are the student who eagerly packed your bookbag and got on the school bus. You treat everyday occurrences and happenings as opportunities to learn and grow.

You love mastering new skills and exploring topics because of the positive feelings you experience in that pursuit. Love of learning is a strength that stimulates cognitive engagement and curiosity. This strength goes beyond the initial curiosity however in one's tendency to systematically accumulate bodies of knowledge. If you are high in this strength, look up from wherever you are reading this book. You may have several degrees or certificates hanging on your wall.

It doesn't matter whether there is an immediate benefit from the learning, the strength propels you forward with persistence and excitement, often in an organized and disciplined manner.

Interestingly, a love of learning correlates most strongly with an engaged, meaningful, and pleasurable life. If you want to increase a sense of connectedness and flow in work, play, and relationships, this is a strength you want to cultivate.

Love of learning is a character strength that supports an open mindset. You are open to exploring new topics, people, and experiences – you are a lifelong learner. Each day is filled with new experiences opening new doors – like Alice in Wonderland. Research demonstrates that love of learning motivates people to persist through negativity, setbacks, and challenges.

Love of Learning and Your Brain

There is a fascinating relationship between our brains and learning. The physical structure of the brain actually changes through learning, which then alters its functional organization. Our brains change throughout our life spans. This is known as neuroplasticity. Learning organizes and reorganizes our brains! This is one of the most impactful medical neuroscience discoveries in 400 years. It has changed the landscape of medicine, education, psychology, and how we see ourselves in relation to our world.

As well, different parts of the brain may be more ready to learn at different times. As brains mature, more fibers grow, and the brain becomes increasingly interconnected and differentiated. These interconnected neural networks connect new and prior learning. This aids in forming memories and promotes academic and social learning. This process, called the associative learning rule, explains how brain cells acquire knowledge (Baysan, 2021).

You may have heard the phrase "Neurons that fire together, wire together." When this happens, learning has taken place. When two neurons fire together they send off simultaneous impulses making the connections between them, the synapses, grow stronger (Peshek, 2019).

Music is a human activity that lights up the whole brain (Sue, *Study suggests listening to music lights up the whole brain*, 2011). In addition to the many cognitive and emotional benefits music provides, one study (Weber, 2013) showed that as little as four years of musical training in youth improved certain brain functions forty years later.

Affirmation

"I will learn something from everyone and every situation."

Quick Tip

Have the heart and eyes of a child: ask "why?" The adult version of this is: tell me more.

Musical Examples of Love of Learning

- Pursuing higher education in music
- Studying multiple interests or instruments beyond your main instrument or focus
- Obtaining certifications in new methods
- Reading the latest research in your field
- Connecting with other professionals to share thoughts or best practices
- Humbleness and the willingness to be a beginner
- Excitement at learning about a musical period or figure
- Willing to try to write in a new style
- Learning to conduct a new piece by a composer
- Playing in a new genre
- Exploring or mastering new music tech or crypto

Discussion Questions/Self-Reflections

Think about your most positive learning experience, no matter what age. Where were you? Who was guiding it? What made it so positive for you? Write down as many of the details as possible.

_____
_____
_____
_____
_____
_____

Treble and Bass

Treble • Overuse of Love of Learning
KNOW-IT-ALL-ISM: you need to have the recognition for being the most knowledgeable one and expect others to defer to your expertise.

- Music marketer who micro-manages their staff rather than let them do their job.
- A&R professional who is closed-minded to music outside of their taste.

Bass • Underuse of Love of Learning

COMPLACENCY: will not go the extra mile to learn something new or crucial to growth. Settles for "this is ok, it's working."

- Educator who yearns to further their education in technology but settles with saying they are "computer illiterate."
- Pop performer who knows they must spend more time on dance training but doesn't make the time for it.

Lights, Camera, Action!

- Tune into a podcast on part of the industry you know nothing about.
- Check out some music scores and dive deeper into a piece of music you already love.
- Take up a new instrument!
- Travel to a new place, even in your city.
- Take different routes and blend education with novelty.
- Look up the history of the names of local streets and businesses.
- Find a friend with interests that differ from yours and have a teaching-learning date.
- Analyze the differences in the reporting style of the same subject from differing news outlets.
- Decide to visit a different museum or historic community building once a month.

Love of Learning VIP

Iannis Xenakis • Music Theorist, Architect, Composer, Engineer

Those who love learning marvel at the veritable font of knowledge each day brings. They delve into subjects some may graze over, they find interest in the deeper realms of understanding, and they often pursue multiple areas of interest, hobbies, or careers. If you can't already tell by Iannis Xenakis' multi-hyphenate status, the man loved learning! A lover of math, Iannis gravitated to multiple expressions of numbers in music theory and electronic music (Kissel, 2022). He created many complicated compositions and authored books on his love of the intersection of math and music. Xenakis' interests encompassed not only a myriad of musical and math topics, but also languages and architecture (*Iannis Xenakis*, 2022). He received a degree in civil engineering while also studying Ancient Greek. His curiosity drove him deeper into all his fields of interest. Later in his life with the advent of computers, he created a computer program that would 'play' his architectural drawings musically.

References

Baysan (2021, September 21). *What is association rule learning?* Medium. Retrieved January 19, 2023, from https://medium.com/codex/what-is-association-rule-learning-abd4a76144d8

Kissel, T. (2022, December 1). *Iannis Xenakis, the Greek composer who revolutionized music.* GreekReporter.com. Retrieved January 29, 2023, from https://greekreporter.com/2022/12/01/iannis-xenakis-greek-composer-who-revolutionized-music/

Peshek, S. (2019, April 30). *How do synapses work?* Texas A&M Today. Retrieved January 19, 2023, from https://today.tamu.edu/2018/01/05/how-do-synapses-work/#:~:text=Synapses%20are%20part%20of%20the,those%20neurons%20to%20the%20muscles

Sue (2011, December 6). *Study suggests listening to music lights up the whole brain.* Therapy Toronto News. Retrieved January 29, 2023, from https://therapytoronto.ca/news/2011/12/study-suggests-listening-to-music-lights-up-the-whole-brain/

Weber, B. (2013, November 10). *Music training in childhood boosts the brain in adulthood.* Medical News Today. Retrieved January 29, 2023, from https://www.medicalnewstoday.com/articles/268597#Music-to-our-ears

Wikimedia Foundation (2022, December 21). *Iannis xenakis.* Wikipedia. Retrieved January 29, 2023, from https://en.wikipedia.org/wiki/Iannis_Xenakis

Perspective

Musical Prelude to Perspective

A long-term career in music is much like moving your little car through a classic board game through different worlds, challenges, and periods of relative luck, sometimes due to a dice toss. Having perspective on the events of your career path is like stepping back from the table to see the whole game and how all the roads connect. It's the refreshing ability to de-invest in momentary outcomes and de-personalize interactions and failures. This may seem quite stoic, but those who exhibit great perspective often only do so after living great amounts of life. They can see that not every rejection is the end of the line and not every win will guarantee future results. They have been "around the block" long enough to anticipate people's behaviors and predict both pitfalls and opportunities. When young musicians have perspective, they might attribute this to witnessing their parents or other family members go through the business in some way. Watching the ebbs and flows of someone else's long-term career path is enlightening and helps to widen the lens and definitions of 'successes and "failures."

Perspective can be employed in many ways. It is for sure your best friend during times of change or decisions. It bravely holds hands with judgment and helps you figure out what may be the best choice played out over a longer stretch of time. For instance, one may have a well-established position at a small school but are offered the chance to have a lesser position at a better, research school that aligns with your future trajectory. It's important to play twenty questions with perspective in this case. Another example is the perceived and actual value of musical output. One band may have tons of streams on Spotify and see pennies to the dollar while another band might have pressed their album on vinyl and generally sell it out at every local show. Depending on your perspective, each scenario has different benefits.

The business side of music is engaged in perspective on every level. Whether one is deciding which artist to put out first, which one to tour, how much money to spend on advertising, and which platforms or spots to spend on, there are endless decisions that rely on the perspective of the past weighed against the hopes of the future. Big artists fund baby artists in publishing and on the label side. This ecosystem is a cycle of curation and growth. There are shifting highs and lows at every level and just like the board game, it's instructive to stand back and take in the larger picture.

Hardships, injuries, pandemics, and losses put things in perspective. As the artistic communities struggled to continue their creativity during the COVID-19 lockdown period, new opportunities and activities emerged. People changed entire methods of operation, adopted technology in unforeseen ways, and discovered their true priorities pretty quickly. This trial by fire forced us all to have a perspective and brought in two other important strengths – creativity and gratitude. Creativity takes on the challenge to find new ways to do old tricks. Gratitude is the helper that allows us to keep a level head and count our blessings when things go wrong, showing us that we already have a full trophy case. Perspective is the slow, intentional dusting of those trophies to remind us of our enduring shine and true values.

What Is Perspective?

Perspective is taking the big picture view; using your wide-angle lens to see things. People who are high in perspective are generally good listeners, ask pertinent questions, are sensitive to context, have a high degree of self-knowledge and can give insightful advice.

Perspective is about being able to see the whole of a situation. You can see the forest and avoid getting tangled in leafy details (*Perspective*, n.d.). Your perspective allows you to stand back and analyze things from a bigger picture and global view. It's like your brain can push the "full-screen view" button. You may be sought out for this sort of advice from others as they trust your ability to think bigger than the problem at hand. You may even have wisdom on broader life events like change, loss, transition, or transformation.

Perspective is distinct from intelligence but represents a high level of wisdom. It is the capacity to give advice and recognize and weigh multiple sides before making decisions. Listening to others with perspective helps you simultaneously think about life lessons, proper conduct, and what's best for the people and situation at hand.

Having perspective is like going up to the highest seat in the theater and looking down at the stage. Viewing from a higher vantage point usually reveals that few things are absolutely right or wrong. You can also peer out into the future with a perspective to see how relatively large or small an issue may be in the long run.

Schwartz and Sharpe say that perspective is essential in having "practical wisdom" which in turn balances out the full symphony of our strengths (Monk, 2021). It is the reasonable voice in our head that maximizes our use of all the strengths while preventing the over or underuse of any one of them.

Research from the VIA indicates perspective is associated with the ability to learn from mistakes and balance the short- and long-term consequences of actions. Individuals high in perspective are valued by others as providing wise counsel.

Perspective and Your Brain

Perspective is the strength that bridges the gap between thinking and feeling. It requires not only cognitive abilities such as weighing evidence and rational decision making but also understanding people, feelings, empathy, behavioral nuances, and insight into one's own limitations.

This is quite a balancing act of many brain functions! Primarily, perspective uses the frontal lobes of the brain responsible for thinking, planning, memory, and judgment. It also involves the parietal lobes primarily responsible for body sensations and touch; the temporal lobe responsible for hearing and language; and the occipital lobe responsible for vision (Stangor & Walinga, 2014). In other words, all aspects of brain functioning are working together while using perspective!

Affirmation

"I enjoy and am sensitive to multiple points of view – they enrich my life."

Quick Tip

Listen deeply before offering your good advice.

Musical Examples of Perspective

- Understanding the effect of culture and time on music history
- Staying grounded through performance errors
- Having a tough skin for bad reviews
- Stepping outside of yourself to analyze a technical problem
- Staying calm through technology glitches and failures
- Maintaining a long-range career plan
- Seeing through and learning from disappointments and setbacks
- Counseling others through their insecurities or stage fright
- Embracing world music and tradition
- Representing multiple genres or types of composers on a concert program

Discussion Questions/Self-Reflections

Think of a time in your life when you needed input from someone wiser and more experienced than you. Who did you choose and why? What was most helpful about their response? How did you apply what they shared with you?

_____

_____

_____

_____

_____

_____

Treble and Bass

Treble • Overuse of Perspective

OVERBEARING: excessive behaviors, ivory tower mentality, arcane pedantic thinking.

- Composer who doesn't allow the musicians' own expression in a performance of their piece.
- Program director who over-manages competent teachers.

Bass • Underuse of Perspective

SHALLOWNESS: foolishness, superficiality, selfishness, makes non-insightful choices or decisions.

- Music photographer who uses poor shots of the artist because they prefer the picture's composition.
- Member of a vocal group who only cares about their own performance and not the blend of the rest of the group.

Lights, Camera, Action!

- Write a song from another person's point of view.
- Research the political and social elements surrounding a piece of music you are performing.
- Assign your class to analyze a rap song that speaks to a cultural or social issue or event.
- Create a new habit from a wise quote.
- Pause, breathe, and connect before responding to someone today before giving your ideas.
- Revisit a challenging experience from your present perspective. Start with "this is what I see now."
- Have an imaginary counseling session with a wise figure, living or dead. Ask questions and imagine their responses.

Perspective VIP
Hiro Murai • Music Videographer

Perspective operates on many levels for music videographers. Not only do they bring their own vision, but they also filter those ideas through the lens of each different artist and their intended message. Then quite literally there are hundreds of options for the actual "perspective" of shooting and editing that story and endless artistic treatments for that vision. Grammy-award-winning music video director Hiro Murai is referred to as a "master of restraint" for his ability to evoke imagery through careful visual perspective (*Hiro Murai*, 2023). He prides himself on his ability to make decisions after carefully evaluating different artistic directions (Wallace, 2018). His perspective is also enhanced by his Tokyo roots and Los Angeles upbringing and feeling out of place. He has applied his avant-garde perspective to music videos, films, and television episodes. https://www.gq.com/story/hiro-murai-breakout-profile-2018

References

Monk, S. (2021, May 21). *The strength of perspective*. The Positive Psychology People. Retrieved January 29, 2023, from https://www.thepositivepsychologypeople.com/the-strength-of-perspective/

VIA Institute on Character (n.d.). *Perspective*. Retrieved January 29, 2023, from https://www.viacharacter.org/character-strengths/perspective

Stangor, C., & Walinga, J. (2014, October 17). *4.2 our brains control our thoughts, feelings, and behaviour*. Introduction to Psychology 1st Canadian Edition. Retrieved January 29, 2023, from https://opentextbc.ca/introductiontopsychology/chapter/3-2-our-brains-control-our-thoughts-feelings-and-behavior/

Wallace, A. (2018, November 30). *'Atlanta' director Hiro Murai has the Vision*. GQ. Retrieved January 29, 2023, from https://www.gq.com/story/hiro-murai-breakout-profile-2018

Wikimedia Foundation (2023, January 24). *Hiro Murai*. Wikipedia. Retrieved January 29, 2023, from https://en.wikipedia.org/wiki/Hiro_Murai

Judgment

Musical Prelude to Judgment

A musician's brain is finely tuned to judgment. Upon hearing music, we are already forming a judgment about whether we find it pleasant or aligned with our tastes. When we analyze music, our ears decide which chord we hear. When we listen to people playing songs on the instrument we play, we consciously and subconsciously make judgments on their ability, which leads to comparisons between us and them. Musicians decide which genre suits them best, which repertoire is the right choice for the audition, and whether to take the gig. The ecosystem of music itself demonstrates judgment by way of orchestral chair ranking, the number of streams and view counts, chart placement, and Grammy awards.

There are also constant financial judgments that accompany decisions like signing artists, releasing songs, and all aspects of publicity. Teachers form assessments to make judgments on the success of learning outcomes and students often compare themselves to the abilities and talents of their peers. Music therapists make crucial judgments in care to make sure their patients meet their goals. All these decisions are formed via a complex dance of metrics, assessments, networks, talents, and true and perceived value.

All this judgment is helped by perspective, experience, and honesty. Judgment is the careful dissection of fundamental facts. Perspective is the 100-foot view that helps us reassemble those facts to make a cohesive and informed decision. Honesty forces us to question our own bias. And experience is the "Judge Judy" in our mind who has seen it all, helping us form wise opinions.

It's important to discern between being judgmental and having judgment. They are different. When someone is judgmental, they are forcing their own values on another system or person. When they show judgment, they are weighing points of view from a non-biased stance to make an informed decision. A judgmental person has already decided the verdict as you walk into their courtroom. A judge with good judgment will listen carefully to the facts provided and make an expert, non-biased decision.

It's important to have judgment as it correlates with our value system. We may recognize something is brilliant but not want to replicate it ourselves. It wouldn't be the right fit for us. Many artists make conscious decisions on how they want to be represented that go against the norms of expression. Think of the heroes who broke the boundaries and allowed more freedom in the expression of body types, race, ethnicity, gender, sexual identities, and age. They made a self-judgment that it would be worth the risk to be themselves. No doubt, someone on their team likely advised them against being so bold and so different for whichever mold they were about to break. These risks can reap great rewards or can also bear great consequences. It is important to ground yourself in your own values when making decisions about the way you want to portray your artistic self in the culture. Regret usually doesn't come from going with one's gut but from betraying it.

Having good judgment is a gift. It's seeing both sides of a situation or problem and having the ability to turn down the radio static of bias and listen without prejudgment or prejudice. Developing personal balanced judgment will help in all aspects of your music career.

What Is Judgment?

Judgment is about discernment - looking at things from all sides. You are using judgment when you think things through with critical thinking skills. One might also call you open-minded. You rely on solid evidence to make decisions rather than jumping to conclusions. You see things in the 360-degree round and can change your mind when differing evidence shows up. The VIA Survey uses the word judgment differently from what one might expect. They do not mean the common word "judgmental"; to look down on or criticize someone. Here, judgment is the very intellectually disciplined process of actively, skillfully conceptualizing, applying, analyzing, and synthesizing your experiences.

Judgment is a "strength of the head" (*Judgment*, n.d.) and is a very thinking-oriented wisdom character strength. We can think of it as the "Tevye Strength." In his famous song, the wise Jewish father from Fiddler on the Roof ponders the way his life would differ if he had money. Research shows that the strength of judgment allows people to see things from many different perspectives which is especially useful in times of transition and change (*Future proof: Solving the 'adaptability paradox' for the long term*, 2021). Think of the many ways COVID has strengthened our judgment and adaptability, especially in the arts.

Cultivating the strength of judgment contributes to more accurate decision making by counteracting biased thinking and providing stability and balance. A discerning, flexible, open mind keeps you fresh and engaged. Judgment is a corrective strength that counteracts faulty thinking like favoring a dominant viewpoint without further analysis of a less-dominant view. When you exhibit judgment, you don't simply cast aside evidence because it conflicts with your currently held beliefs. This strength challenges you to weigh things based on evidence, even if it means a change of plans or goals.

Judgment and Your Brain

The kingdom of critical thinking in our brain is the prefrontal cortex (PFC). This is the seat of executive function control centers (Willis, *Understanding how the brain thinks*, 2011). The PFC is the reflective "higher brain" as compared to the reactive "lower brain." It gives us the potential to have voluntary control of our thinking, emotional responses, and behavior. It also comprises 20% of our brain volume which is the highest percentage of any brain structure (Willis, *Improving executive function: Teaching challenges and opportunities*, 2011). Animals are more dependent on their lower reactive brains to survive. Humans developed with the luxury of a larger reflective brain. The PFC was the last part of the human brain to mature through evolution.

This maturation is a process of neuroplasticity that includes the pruning of unused cells to better provide for the needs of more frequently used neurons. This neuroplasticity also strengthens the connections in circuits that are most used growing stronger and more connections among neurons. This leads to the application, analysis, and synthesis of critical thinking and judgment.

Affirmation

"I constantly grow by having an open mind and seeing the world and people through new eyes. I weigh all aspects of my decisions objectively – even arguments that conflict with my beliefs."

Quick Tip

Be discerning – the devil is in the details. Keep turning the picture frame around to see it from every angle.

Musical Examples of Judgment

- Providing balanced performance feedback to a student or contestant
- Counseling someone on artistic choices or career paths
- Allocating accurate and appropriate funding for a program or event
- Executive ability
- Trustworthy amongst peers to make fair decisions
- Neutral self-assessment
- Good sense of the relative importance of a mistake or slip-up
- Listening to both sides of a student/teacher complaint without bias

Discussion Questions/Self-Reflections

Think of someone in your life that you go to for advice or "critical thinking." How have they helped you decide things in the past? Break it into specific behavior steps and then apply it to another situation you are currently weighing.

_____
_____
_____
_____
_____
_____

Treble and Bass

Treble • Overuse of Judgment
CYNICISM: skepticism. I overanalyze everything, breaking it apart to the point of annoying others and often ruining the fun and joy of "the whole gestalt" of experiences.

- Songwriter who doesn't send out releases to playlists or blogs because they believe it won't matter.
- Entrepreneur who holds back from promoting his/her business because of negative bias about self-promotion.

Bass • Underuse of Judgment
UNREFLECTIVENESS: I speak without thinking or filtering, valuing my spontaneity and honesty above all else.

- Middle school music teacher who doesn't realize the racist undertones of a particular song.
- Voice teacher who criticizes the range of a song as being "inappropriate" for a transgendered student

Lights, Camera, Action!
- Create a ten-song playlist based on a prompt (music for driving, music for Italian cooking) and include the best songs you can find.

- Follow the proceedings of a music copyright case and consider all points of view.
- Select a news station or a political program that is an opposite point of view from your own thinking. Keep an open mind and summarize three main points you heard.
- When you meet someone with a different approach or set of beliefs, ask a few clarifying questions to better understand their point of view.

Judgment VIP
Nadia Boulanger • Music Teacher

Teachers spend their entire day making judgments and choices to be effective and best do their job. From lesson plans to real-time snap judgments and pivots in approach, to grading – judgment is an integral part of being a teacher. Fierce female composer and pedagogue, Nadia Boulanger (*Nadia Boulanger*, 2023) is known as one of the most famous composition teachers in history, with a student roster that boasts the likes of Aaron Copeland, Astor Piazzolla, Phillip Glass, Elliot Carter, and Quincy Jones. Not only did she help guide these legends with their own music, but she was also able to carefully exist as a very outnumbered woman in a very male sphere. Her judgment and perceptiveness guided her through this terrain all the way to the conductor's pulpit where she was the first female conductor of the New York Philharmonic and the Boston Symphony Orchestra. Nadia's choice to navigate through these halls while diaphanously staying behind the scenes allowed her into the backstage of areas not open to women in music at the time (Robin, 2021).

References

VIA Institute on Character (n.d.). *Judgment*. Retrieved January 29, 2023, from https://www.viacharacter.org/character-strengths/judgment-critical-thinking

McKinsey & Company (2021, August 13). *Future proof: Solving the 'adaptability paradox' for the long term*. McKinsey & Company. Retrieved January 29, 2023, from https://www.mckinsey.com/capabilities/people-and-organizational-performance/our-insights/future-proof-solving-the-adaptability-paradox-for-the-long-term

Robin, W. (2021, July 30). *She was music's greatest teacher. and much more*. The New York Times. Retrieved January 29, 2023, from https://www.nytimes.com/2021/07/30/arts/music/nadia-boulanger-bard-music.html

Wikimedia Foundation (2023, January 24). *Nadia Boulanger*. Wikipedia. Retrieved January 29, 2023, from https://en.wikipedia.org/wiki/Nadia_Boulanger

Willis, J. (2011, June 13). *Understanding how the brain thinks*. Edutopia. Retrieved January 29, 2023, from https://www.edutopia.org/blog/understanding-how-the-brain-thinks-judy-willis-md

Willis, J. (2011, September 1). *Improving executive function: Teaching challenges and opportunities*. Edutopia. Retrieved January 29, 2023, from https://www.edutopia.org/blog/improving-executive-function-judy-willis-md

THE TRANSCENDENT STRENGTHS

Appreciation of Beauty and Excellence

Musical Prelude to Appreciation of Beauty and Excellence

The importance of beauty and excellence, especially in the arts, cannot be overvalued. It is essential to the core of the artistic impulse – to commemorate, recreate, and celebrate life's allure and also, its anguish. An artist is compelled to abstract experience into something eloquently emotional, inspiring, or heartbreaking. To exquisitely crystalize a real universal moment – whether in a lyric, an Olympic ice-skating routine, a song, or an orchestral movie soundtrack – is breathtaking. The tear that rolls down your cheek is a testament to that wonder. Sometimes these moments are described as undeniable. When you witness something so beautiful or excellent, you just know, and you simply feel. It's as if somewhere deep inside a bell is stuck and you ring with truth.

As an artistic community, we have an extra sensitivity to not just the product but the process of creating. We understand the literal blood, sweat, tears, time, money, frustration, elation, and sense of accomplishment and birth inherent in making something original, beautiful, and hopefully enduring. Creation in all forms is holy and taking time to sit 'in awe' is an act of prayer.

Music is a touchpoint for this strength. It is an act of beauty and excellence for those who participate in it and it's an experience for those who witness it. Hearing music that captivates us is the beginning of a passion. It awakens something eternal in us. From that initial encounter, one might become a fan. Others may be inspired to create and take up music as a very literal part of their daily identity. And anyone who has fallen in love with a band, musician, song, musical, or composer can attest to the deep connection that is felt. It is almost like falling in love. When you are in the presence of something so real to your soul, you immediately feel seen. And you yearn to spread and share this love with as many people as you can. This is how superfans evangelize the latest album, music video, or live show of their favorite artist. They aggregate into the little monsters, hives, navies, and Beliebers and spread their love via their own actions, their words, and their social platforms. Fans are capable of such amazing motion in a musician's career. They bear witness to the excellence and return for that experience over and over again.

Gratitude and mindfulness are deeply woven into appreciating all things beautiful and excellent. Those who take the time to absorb their thankfulness profoundly take in joy. They partner with the delight of what they are seeing, hearing, tasting, smelling, touching, and experiencing. In other words, their noses are already buried in the petals of life's roses. And those roses are not at all passing to them – they are the main event. Time moves at a slightly slower pace for those who really take in the beauty around them.

And there is also a special place for those who appreciate the arts by donating their money, time, and influence to help preserve, fund, and uplift.

But regardless of economic stature, we all can bring attention to the glory of music. Whether it's the grandmother who sings songs from her culture to pass down timeless wisdom, the violinist who plays a memorial to someone beloved, or the composer who so perfectly captures a fleeting instant in permanent sound, we can all serve and receive the cycle of beauty and excellence and cherish it for time immortal.

What Is Appreciation of Beauty and Excellence?

Appreciation of beauty and excellence is one of the transcendent strengths that provide meaning and forge connections to the larger universe. Beauty, excellence, and skilled performance are noticed in all domains of life, from science, to art, to mathematics, and even everyday experiences. Appreciation of beauty and excellence (ABE) elicits elevated positive feelings such as awe and wonder.

Simply put, ABE is about seeing the best in people, experiences, and things around us. It opens you up to be totally alive and in the present.

This is a feeling strength and has a strong set of emotions connected to it. If you possess this strength, you don't think about it cognitively; you feel it. ABE activates the senses: sight, hearing, touch, smell, and taste. You may frequently say things like, "look at that sunset" or "I wonder who wrote this magnificent piece of music." You experience a sense of awe and wonder as you experience your world.

Appreciating beauty feels as if you have your own internal coloring book with cinematic sound effects. You feel your world is bright and you don't miss the small stuff! The transcendent power of ABE allows you to move out of whatever reality you're experiencing into a world of eternal beauty and excellence. You experience a larger perspective that connects you to the universal, meaningful, and eternal aspects of life.

The value of this strength is that it creates positive emotional experiences that help people cope with difficulties, pain, and loss. When struggling with negative feelings in a relationship, ABE helps you to shift; to focus on what is good and uniquely beautiful about this person. It allows you to notice excellence in everyday situations – a server, a grocery store clerk, or someone on your team. Columnist and author (and Nancy's son) Scott Kirsner described this as a gift, "When I get a chance to pick up a trick for nothing – to watch a professional at work, tuition-free, I can appreciate what a deal that is. Because ordinarily, you know, you've got to pay for an education" (*Scott Kirsner: Home*, n.d.).

Appreciation of Beauty and Excellence and Your Brain

Because of its non-definite nature, science has shied away from the question of beauty and excellence. However, in the last twenty years, positive psychology has helped form a science of happiness by studying the effect and scope of life pleasures such as the appreciation of art and beauty, and the positive emotions: happiness, calm, gratitude, awe, joy, and love.

Today, the field of "neuroaesthetics," represents a merging of neuroscience, art, ABE, and social interaction (Pak & Reichsman, 2017). This has added science about the neural mechanisms that have provided a more biological and evolutionary lens to help us understand the ABE.

It is unlikely that there are brain systems specific to the appreciation of art/music and excellence. Most likely there are general aesthetic systems that determine how appealing an object is, be it a donut or a piece of music. Neuroaesthetics seeks to identify brain areas that specifically mediate the aesthetic appreciation of artworks. This is done through neuroimaging studies of vision, hearing, taste, and smell.

Studies show that the most important part of the brain for aesthetic appraisal is the anterior insula; deep within the folds of the cerebral cortex. This is the part of the brain that responds when we see something beautiful or excellent (Brown & Gao, 2011).

Researchers have identified certain body responses and facial expressions such as wide-open eyes, an open mouth, goosebumps, tears, and a lump in the throat that typically accompany beauty and excellence experiences (Ekman & Keltner, 1997). It was also discovered that after an elevating experience of beauty and excellence, a sense of grateful admiration wells up (Emmons & McCullough, 2003).

In addition to things like music, art, architecture, sports, and nature, ABE is often connected to religious and spiritual experiences. This strength is often a pathway for moral and spiritual advancement. Experiences of awe and wonder can be intimately connected with a sense of the power of the divine. Evidence of this strength can show up as profound gratitude for both the beauties of creation and the powers of the natural world. Notice the way you may feel after walking out of a concert space after being immersed in music that inspires you.

Affirmation

"I find beauty all around me – in people, nature, art, ideas – and in the simple things of life."

Quick Tip

Stop and smell the roses in your life.
Turn off your thinking brain and let your senses be your guide.
See your world through the awe of the child – with eyes wide open.

Musical Examples of Appreciation of Beauty and Excellence

• Feeling awe or crying at someone's skilled audition or perfect performance

- Giving compliments to students freely
- Decorating your teaching studio to be warm, positive, and inviting
- Paying great attention to stage wear
- Insisting your choir learns choreography to enhance their performance
- Seeking out studio equipment for aesthetic value as well as function
- Spending time picking out the best photographer or videographer
- "Fanning out" for your favorite band in a review of their concert
- Rewarding top employees of the label

Discussion Questions/Self-Reflections

At its core, ABE is about awareness of uniqueness, virtue, skill, and exceptionality across many domains in life. Generally, there are three different ways this tends to show up: in nature, skills, and talents, and through the goodness and virtues of people. Given these categories, describe some of the beauty and excellence you are appreciative of in your life.

_____
_____
_____
_____
_____
_____

Treble and Bass

Treble • Overuse of ABE
PERFECTIONISM: an intolerance of others without this strength, snobbery, obsession with details to the point of stifling flow.

- Artist who won't release anything because it's not up to par.
- Entrepreneur who misses a venture capitalist funding opportunity because they are perfecting their pitch.

Bass • Underuse of ABE
OBLIVION: lose sight of what is happening in the present moments, unconsciousness.

- Member of a string quartet who only pays attention to their own part.
- Vocal coach who doesn't notice a singer's emotional portrayal of a role and focuses only on the technical.

Lights, Camera, Action!
- Redo your socials or website to truly reflect the vibe of your musical aesthetic.
- Decorate your office to make clients feel more relaxed, open, and creative.

- Go through each of your senses with a slow lens camera and record what you see, smell, taste, hear, and feel.
- Go outside and find five things you find beautiful or represent excellence for you.
- Listen to your favorite song or a piece of music and actively notice what you appreciate about it.

Appreciation of Beauty and Excellence VIP
Steve Aoki • DJ • EDM Artist • Producer

Steve Aoki has built an entire career following his love of what he does, not just musically but connecting to fans world-wide. He speaks often in interviews about pursuing excellence not only with his incredible work ethic but in his physical and mental health (King, 2022). His Aoki Foundation focuses on research in brain health and diseases such as Parkinson's, Alzheimer's, and dementia (Yopko, 2022).

He speaks about making choices from his gut. He trusts his own emotional response to work and stresses the importance of mindfulness and gratitude. Listening to Steve Aoki speak about his passions, one cannot help but notice his growth mindset. He is expansive in the ideas he pursues professionally and his many points of interest. He is constantly challenging his own excellence and striving to be better in all categories of his personal and professional life. Paired with his humanity and gratitude, Aoki is a powerful demonstration of a resilient and energized artist who seems to only get better and more inspired with age and experience.

References

Brown, S., & Gao, X. (2011, September 27). *The neuroscience of beauty*. Scientific American. Retrieved January 29, 2023, from https://www.scientificamerican.com/article/the-neuroscience-of-beauty/

Ekman, P., & Keltner, D. (1997). Universal facial expressions of emotion: An old controversy and new findings. In U. C. Segerstråle & P. Molnár (Eds.), *Nonverbal communication: Where nature meets culture* (pp. 27–46). Lawrence Erlbaum Associates, Inc.

Emmons, R. A., & McCullough, M. E. (2003). Counting blessings versus burdens: An experimental investigation of gratitude and subjective well-being in daily life. *Journal of Personality and Social Psychology, 84*(2), 377–389. 10.1037/0022-3514.84.2.377

King, S. (2022, March 10). *How Steve Aoki still conquers the world stage after 20 years of touring*. Forbes. Retrieved January 29, 2023, from https://www.forbes.com/sites/scottking/2022/03/09/how-steve-aoki-still-conquers-the-world-stage-after-20-years-of-touring/?sh=5415e63854fd

Pak, F. A., & Reichsman, E. B. (2017, November 10). *Beauty and the brain: The emerging field of neuroaesthetics: Arts: The Harvard crimson*. Arts | The Harvard Crimson. Retrieved January 29, 2023, from https://www.thecrimson.com/article/2017/11/10/neuroaesthetics-cover/

Scott Kirsner: Home (n.d.). Retrieved January 29, 2023, from http://www.scottkirsner.com/

Yopko, N. (2022, January 21). *Watch Steve Aoki and a neurosurgeon discuss new ultrasound helmet to treat brain disease*. EDM.com – The Latest Electronic Dance Music News, Reviews & Artists. Retrieved January 29, 2023, from https://edm.com/lifestyle/steve-aoki-and-neurosurgeon-discuss-ultrasound-helmet-to-treat-brain-disease

Humor

Musical Prelude to Humor

Humor is a magic elixir for getting things done. It works in almost any situation, within reason. The ability to make someone smile or laugh or take things less seriously is a much-needed break from the heaviness and consequences of life and work. People enjoy working with those who make them feel lighter, happier, and generally less nervous to make a mistake. It brings an air of easy spirit to tasks that rub off on the work itself. It's a gift to work with someone who has good humor. The time may seem to fly by with that person and before you know it, the job is done! Balancing humor with seriousness is important or else things can disintegrate into a laughing fest and nothing gets done.

Humor takes high social intelligence. It's important to be able to get a feel for the room to know what jokes will land or work. There is nothing more cringy than someone who can't read the crowd but continues with unfunny or offensive humor that sinks like lead and is met with awkward silence. Because of how nuanced we all are, it's difficult to create humor that works broadly. Those who can do it are quite skilled indeed!

Humor is a great antidote to nervousness and performance and social anxiety. In addition to lightening the atmosphere, it's a coping mechanism that brings out common humanity. Humor loosens that long walk out onto the stage or to the table on a first date. When done well, humor is an "ice breaker" that brings immediate warmth to a new or uncomfortable situation.

There is also a lot of space for creation within humor. Think of all the musical fun that happens on shows like *Saturday Night Live*. Musical satire is a genre within itself and there are artists who only make comical music, sometimes to express very serious societal problems.

If you are someone who values and uses humor, you most likely feel connected to others who do so. Perhaps this is a major staple in forming teams with people or feeling like you can easily get along with others. Fast-paced music jobs that allow little margin for error need a dose of humor to release some of that pressure. Like any strength, this one must find its happy medium. It can really transform a room, just like how music playing in the background can paint the whole room with a vibe.

What Is Humor?

Humor is a transcendent strength that connects us to others and helps us find levity in the tribulations life provides. Humor lights up social interactions and is a valuable method of coping with stressful situations. It's sometimes typecast as laughing, telling, and remembering jokes, and being outwardly playful. These aspects are all valid but there is a quieter side to humor – seeing the lighter side of things, storytelling that brings smiles to others, and having a playful spirit and approach to experiences and people. Whichever ways you choose to express your humor, playfulness, and not taking yourself too seriously is at its foundation.

Humor and laughter are sometimes called the body's natural opiates and antidepressants. They relax us, improve our mood, and have many physical and emotional benefits. Being around people you laugh with is an intoxicating natural high. When you find yourself taking things too seriously, finding humor in the situation is helpful.

As with all the strengths, you should find the balance point – the golden mean. It's easy to overuse humor to avoid problems and serious issues in life. Too much humor can also be read as uncaring and can be off-putting or hurtful in relationships.

People tend to connect to others by way of humor. It is a great bonding moment to share a favorite comedic sitcom or to experience something hysterical with someone. People who connect over funny things "are likely to rate higher on the traits of extraversion, agreeableness, and openness to experience" (*Humor*, n.d.). People who can work together with this lightness of spirit can feel they are in flow together. Time seems to go by uncounted and tasks get done without the heaviness they might otherwise carry.

Not only can humor connect us, but it can defend us as well. Sigmund Freud noted that humor "allows us to maintain a healthy perspective in the face of misfortune and adversity" (Serani, 2022). Ask any comedian where they may have gotten their sense of humor, and you will probably hear how they used it as a tool to overcome some hardship or as a defense mechanism. Also, notice the many parody songs that exist as vehicles to protest and bring awareness to injustices and other very serious issues.

Humor and Your Brain

There are two structures in our brains that cause us to laugh. Both are in the limbic system: the amygdala and the hippocampus (Tuarez, 2021). Laughter in the human brain is a social response that boosts our feel-good chemicals serotonin and dopamine. Laughter appears to be a crucial development in social interaction and communication, existing across all human cultures (Raine, 2022). Even rats and chimps laugh! And if you've ever watched a baby develop, you know that at around the age of six months, they begin to develop a sense of humor by smiling and laughing to stimuli (Michel, 2017).

There is truth to the saying "laughter is the best medicine." It has psychological, physiological, and emotional benefits. Laughing with others releases our feel-good chemicals, endorphins, in our brain via opioid receptors. This social laughter has been shown to increase long-term bonding and relationships (Manninen et al., 2017). It is also quite contagious! One member of your group's laughter can start a chain reaction, thus triggering an endorphin release in the whole group (*Social Laughter releases endorphins in the brain*, 2017). Studies have also shown that laughter has a similar effect to antidepressants in the way it activates the release of serotonin (Cha & Hong, 2015). It does everything from increasing mental health (Mora-Ripoll, 2010) to protecting your heart (*Laughter is good for your heart*, 2000)!

Humor also may help us complete daunting and tedious tasks, according to a study conducted by Australian National University management professors David Cheng and Lu Wang (Michel, 2017). And how often is our funny bone tickled? A University of Ontario study showed that adults laugh an average of 18 times a day, and children tend to laugh three to five times more (Tuarez, *What part of the brain controls laughter?*, 2021).

Affirmation

"I approach my life playfully – and look for humor in difficult and stressful situations."

Quick Tip

Humor enhances pleasure and positive emotions in life, all contributing to overall happiness. So, laugh a lot, alone, and with others!
Laughter is free!

Musical Examples of Humor

- Telling jokes as an awards show host
- Ability to laugh at recording studio mistakes
- Bringing light and levity to situations of complexity, loss, or grief
- Cheering students up after difficult losses
- Laughing at one's own musical slip-ups
- Using funny analogies in teaching to make a subject more memorable
- Encouraging light-hearted thinking in abstract music analysis

Discussion Questions/Self-Reflections

Of all your relationships, with whom do you laugh the most? How would you describe that person's humor style? How does humor impact that relationship? When do you notice you want to be with that person?

_____

_____

_____

_____

_____

_____

Treble and Bass

Treble • Overuse of Humor
GIDDINESS: overly excited, laughing at everything, buffoonery.

- Classical singer who doesn't take learning language seriously and laughs at foreign-word lyrics.
- Producer who allows the studio to get filled with people who are kidding around too much, and the workflow is derailed.

Bass • Underuse of Humor
OVER-SERIOUSNESS: takes everything at face value and refuses/is unable to find lightness, intolerance for others' humor.

- Audio engineer over-protecting expensive equipment to the point that musicians feel uncomfortable and stiff.
- Choreographer who intimidates the performer by going too fast for their learning curve in the name of "working at a high level/fast pace."

Lights, Camera, Action!

- Change the words of a song to humorously reflect something going on at work.
- Make your next get-together a costume party with a theme and encourage everyone attending to stay in character.
- Laugh more and spend time with people who make you laugh.
- Tell jokes and listen to other people's jokes.
- Go to some great stand-up comedy near you.
- Find comedy podcasts and share them with friends.
- Try a YouTube video on laughter yoga (it's a real thing!).
- Think of a way you could bring playfulness or fun to a difficult situation or relationship in your life.

Humor VIP

Bo Burnham • Comedian • Musician • Actor

Rising to fame after being discovered on YouTube, Robert Picerking 'Bo' Burnham has made a name for himself both in comedy and music. Bo has been very open about mental health issues in his career and has even written and directed *Eighth Grade* which focuses on the anxiety that social media has spurred in its young users. Bo is an example of someone who uses his humor in every way one could use a strength – as a salve, to process the world, as a means of connecting to others and profoundly enmeshed with music to canonize difficult feelings. There was perhaps no better example of this than his pandemic-era one-man-show *Inside* (*Boburnham.com*, n.d.). This raw look at lockdown isolation and depression brimmed with the ache of honesty delivered through darkly funny songs. Using humor as the ultimate tool to process something unfathomable, Bo expressed sentiments his generation could relate to as well as universal longings for a deeper human connection.

References

Boburnham.com (n.d.). *Boburnham.com*. Retrieved January 29, 2023, from http://www. boburnham.com/

Cha, M. Y., & Hong, H. S. (2015). Effect and path analysis of laughter therapy on serotonin, depression and quality of life in middle-aged women. *Journal of Korean Academy of Nursing, 45*(2), 221. 10.4040/jkan.2015.45.2.221

Manninen, S., Tuominen, L., Dunbar, R. I., Karjalainen, T., Hirvonen, J., Arponen, E., Hari, R., Jääskeläinen, I. P., Sams, M., & Nummenmaa, L. (2017). Social laughter triggers endogenous opioid release in humans. *The Journal of Neuroscience, 37*(25), 6125–6131. 10.1523/jneurosci.0688-16.2017

Michel, A. (2017, March 31). *The science of humor is no laughing matter*. Association for Psychological Science – APS. Retrieved January 29, 2023, from https://www. psychologicalscience.org/observer/the-science-of-humor-is-no-laughing-matter

Michel, A. (2017, March 31). *The science of humor is no laughing matter*. Association for Psychological Science – APS. Retrieved January 29, 2023, from https://www. psychologicalscience.org/observer/the-science-of-humor-is-no-laughing-matter

Mora-Ripoll, R. (2010 Nov-Dec). The therapeutic value of laughter in medicine. *Altern Ther Health Med, 16*(6), 56–64. PMID: 21280463.

Raine, J. (2022, September 13). *The evolutionary origins of laughter are rooted more in survival than enjoyment*. The Conversation. Retrieved January 29, 2023, from https:// theconversation.com/the-evolutionary-origins-of-laughter-are-rooted-more-in-survival-than-enjoyment-57750

ScienceDaily (2000, November 17). *Laughter is good for your heart, according to a new University of Maryland Medical Center Study*. ScienceDaily. Retrieved January 29, 2023, from https://www.sciencedaily.com/releases/2000/11/001116080726.htm

ScienceDaily (2017, June 1). *Social Laughter releases endorphins in the brain*. ScienceDaily. Retrieved January 29, 2023, from https://www.sciencedaily.com/releases/2017/06/170601124121.htm

Serani, D. (2022, April 25). *The benefits of humor*. Psychology Today. Retrieved January 29, 2023, from https://www.psychologytoday.com/us/blog/two-takes-depression/202204/the-benefits-humor

Sussex Publishers (n.d.). *Humor*. Psychology Today. Retrieved January 29, 2023, from https://www.psychologytoday.com/us/basics/humor

Tuarez, J. (2021, February 19). *What part of the brain controls laughter?* NeuroTray. Retrieved January 29, 2023, from https://neurotray.com/what-part-of-the-brain-controls-laughter/

Tuarez, J. (2021, January 28). *What part of the brain controls laughter?* NeuroTray. Retrieved January 29, 2023, from https://neurotray.com/what-part-of-the-brain-controls-laughter/

Hope

Musical Prelude to Hope

Music provides a beacon of hope for so many. It can unify us in times of joy and absolute sorrow. It elevates the best moments and brings light and unity to the lowest ones. Music can evoke possibility and power us through the many adversities of life. When we participate in multi-person music-making, we braid our dreams together in its fabric. People playing in a band or orchestra together is a powerful, coordinated act inspiring hope. Singing together in a choir provides audible and visible support and strength to our literal and figurative voices. The many people and parts to putting out a music video require concerted effort driven by vision and hope. Even literal songs of worship and inspiration are designed to act as a spiritual salve and move us into a place of higher optimism and grace. Music may be one of hope's most powerful vehicles.

Hope is the engine that keeps us moving towards something better, higher, or more ideal. We are motivated to keep on going in our careers toward greater heights driven by the hope that we can get there. Musicians are no strangers to how difficult the road is between the dream and reality. Without hope, there seems to be little incentive to put in that much continuous ardor. Hope fuels our work ethic and compels us to go that extra mile, like an adrenaline rush toward the end of a race. There is also an uplifting quality to hope which makes the work feel less laden.

And what is a hopeful journey without a villain, a threat, a storm, a beast, or a downfall? The quality of your faith will be evidenced by your ultimate reactions to obstacles. Will that seventh audition rejection topple your desire to perform? Will the label dropping you make you change careers out of music? Will being downsized from your company make you doubt your own talents and potential? Hope gives us the wings to fly out of the rubble and assess the damage from above. It's a perspective that these hindrances are temporary and are probably here to teach you something. The ability to process trauma with a growth response is perhaps one of the highest aims of hope. When we go through something awful, we are given a free lesson in resilience, self-reliance, and in hope.

Allow the music and hope to intermingle in your life. Add more music to your hope and more hope to your music. Think of the best outcome possible and live right there in that energy. Put your best foot forward always and don't worry, there will always be detractors, people who want to take what you have, and other dangers lingering in the shadows. To allow that energy into your enterprise is to allow yourself to be compromised and poisoned by other people's naysaying. Stay true to your dreams, your hopes, and your spirit. Use your strengths as guidance and protective guardians of your ultimate musical ambitions. Hope, when expressed through music, can be the highest-grade fuel you can put in your tank.

What Is Hope?

Hope is a transcendent character strength that helps us overcome difficult struggles by connecting us to the larger universe through meaning and purpose. Hope is not some idealistic wish or fantasy – it has legs – that is, it has will power and way power. In his pioneering work, *The Psychology of Hope* (Snyder, 1994), Rick Snyder defines willpower as our belief that we can change things – and the courage to try. Way power is the *how* part; the belief that we know how to change things and the belief that it's a solvable problem.

To hope means to have the agency and the pathways to go after the desired goals (*What is hope? why it's important to hope?*, 2022). Positive psychology sees hope in terms of positive future expectations. However, the nature of hope varies from person to person. In positive psychology, hope is a cognitive practice that involves the intentional act of setting goals and working toward them with purpose. It is a driving personal philosophy that believes you can achieve what you want by doing the work and keeping the faith.

Hopeful people set goals and take action to reach them. They also have agency over their ability to take pathways toward their positive future expectations. John Parsi, director of the Hope Center at Arizona State University, says, "Optimistic people see the glass as half full, but hopeful people ask how they can fill the glass full" (Tricoles, 2021). Social science researchers at ASU have found that activating hope is harder work than we may think. They are unpacking the ways in which different populations conceive of hope and exploring how having higher hopes can improve our health and well-being.

Viennese psychiatrist Viktor Frankl (1905–1997) was an example of great transcendence and resilience. He was a prisoner of the Nazis for four years, during which he discovered a way to hang onto hope. His personal journey and trauma resulted in a poignant book *Man's Search For Meaning* (Frankl & Lasch, 1962). Frankl strongly owed his ability to hold onto hope to his inner and core sense of his life's greater purpose. He refers to this as the "will to meaning." When one has a purpose or goals, they can survive great hardships and horrors. The hope of seeing his family again was key to his will to meaning.

More contemporarily, positive psychologist and "hope" teacher Dr. Dan Tomasulo embodies and radiates a fully charged life. He gives active ways to help combat depression using hope (Tomasulo, 2021).

Hope and Your Brain

Hope is transformational for your brain. When experiencing hope, your brain pumps neurochemicals called endorphins and enkephalins which mimic the effects of morphine. These chemicals help the body and brain recover faster and numb or block pain (Small, *Brain bulletin #47 – The Science of Hope*). Some believe hope is as vital to the brain as the oxygen we breathe!

In difficult times, a feeling of hopefulness can make a real difference. Watching movies and listening to stories about hope can have a hugely positive effect. Stories are the strongest brain-state changers on the planet! Neuroscientists believe that our brains are wired for stories and their ability to cathartically let us go along with the characters on their emotional journey. The term used for this is "narrative transport" (Vinney, 2022).

Hope is associated with many positive outcomes: better academic achievement, greater happiness, and even lowered risk of death. Hope is a necessary character strength for meeting everyday goals as well as helping us get through tough times (Weir, 2013).

Affirmation

"I am hopeful, especially when others are not."
"Hope is my beacon of light, showing me paths into my future."

Quick Tip

Maintain positive involvement in life despite imposed limitations.
Activate hope to overcome hurdles and have greater control over your path.

Musical Examples of Hope

- Staying strong during a period of large-scale concert cancellations
- Looking forward to the next performance after a poor one
- Believing in your student's ability to nail that difficult aria in their recital
- Bolstering the spirits of your choir as they go into a stiff musical competition
- Auditioning for American Idol for the eighth time
- Applying to your dream job
- Approaching a major publisher with your portfolio

Discussion Questions/Self-Reflections

Nelson Mandela, the father of the modern nation of South Africa, said, "May your choices reflect your hopes, not your fears" (Mandela, 1990). Can you list some of the things you could attempt to do with a more hopeful outlook? What if you knew you could not fail?

_____
_____
_____
_____
_____
_____

Treble and Bass

Treble • Overuse of Hope
POLLYANAISM: naïve optimism, Panglossian.

- Overstating your draw in a city you've never played to a tour manager.
- Handing in your video game score late and assuming it will be no big deal to the music supervisor.

Bass • Underuse of Hope
NEGATIVISM: immediate assumption of the worst outcome, skeptical of positive outcomes.

- Jazz combo director who forgoes competition submissions because they don't believe the group will rise to the occasion.

- Graduate student who doesn't apply for a grant for their work because they think no one will value it.

Lights, Camera, Action!

- Write down positive words on your score or lead sheet.
- Uplift your class before a big concert by delivering a mindset shifting inspirational speech.
- Think of a struggle or problem you are having. Write down two hopeful, optimistic, and realistic options/thoughts that comfort you.
- Find and watch a movie that expresses a message of hope. Take note of the actions of the characters that led them to hope. See how you can apply this message to your life.
- Take small steps – what can you do today?

Hope VIP
India Aire • Artist • Lyricist

India Aire is an artist whose lyrics aim to elevate people and urge them to accept themselves as they are. Listen to her song "There's Hope" to hear a testament to the importance of keeping hope and gratitude active and present in everyday life. She demonstrates perspective in these lyrics, discussing a conversation she had with a blind Brazilian boy living in poverty who described his life as "paradise." When he asked her what life was like in the U.S.A., all she did was complain. The message in this was how joy is a state of mind and there is always hope to improve your state. As an artist, it is often a risky move to dedicate your music to telling the truth and putting forth a positive message. She is a powerful inspiration to her fans and continues to grow and show her growth in her honest lyrics (YouTube, 2013).

References

Frankl, V. E., & Lasch, H. (1962). *Man's search for meaning: An introduction to lo-gotheraphy.* Hodder and Stoughton.

Mandela, Nelson, 1918–2013. (1990). *Nelson Mandela: the struggle is my life: his speeches and writings brought together with historical documents and accounts of Mandela in prison by fellow-prisoners.* London: IDAF Publications.

Small, T. (n.d.). Brain bulletin #47 – The Science of Hope. Retrieved January 30, 2023, from https://www.terrysmall.com/blog/brain-bulletin-47-the-science-of-hope

Snyder, C. R. (1994). *The psychology of hope: You can get there from here.* The Free Press.

Tomasulo, D. (2021). *Learned hopefulness: The power of positivity to overcome depression.* READHOWYOUWANT.

Tricoles, R. (2021, June 15). *The science of hope: More than wishful thinking.* ASU News. Retrieved January 29, 2023, from https://news.asu.edu/20210615-solutions-science-hope-more-wishful-thinking

Vinney, C. (2022, January 31). *What is narrative transportation?* Verywell Mind. Retrieved January 19, 2023, from https://www.verywellmind.com/what-is-narrative-transportation-5217042

Weir, K. (2013, October). *Mission impossible*. Monitor on Psychology. Retrieved January 29, 2023, from https://www.apa.org/monitor/2013/10/mission-impossible#:~:text= Hope%20is%20associated%20with%20many,also%20for%20meeting%20everyday %20goals.

The Happiness Blog (2022, December 27). *What is hope? why it's important to hope? [psychology]*. Retrieved January 29, 2023, from https://happyproject.in/how-to-hope/

YouTube (2013). *India.Aire and Oprah go soul to soul.* YouTube. Retrieved January 29, 2023, from https://www.youtube.com/watch?v=QxD-hSB7XuY.

Spirituality

Musical Prelude to Spirituality

Music is a bridge for spiritual transcendence. You would be hard-pressed to find any culture during any period that did not employ music as an essential means of expression, ritual, or commemoration. Think about our own cultures and societies in the modern day. Music is ubiquitous as you move through your day in stores, elevators, restaurants, cars, and offices. Music fills places of worship for services, weddings, funerals, baptisms, and all other holy rituals. It is a way for us to connect with each other and have a common human experience. We fall in love to music and share songs with our friend tribes. We teach songs generationally and pass down traditions. Babies learn a tremendous amount of new words, facts, letters, and numbers set to music. We feel meaning in pieces of music that wordlessly communicate by tone only. Music may be one of those truly universal experiences that words can only dimly capture.

Spirituality inspires people to write music that then animates others into good action. Think of the elation and emotionality of being uplifted by an amazingly powerful gospel singer or choir. Or think of the heartfelt story-telling in many other secular or sacred songs designed to resonate with our common struggles. There is such a profusion of emotions and thoughts to express about one's spiritual convictions that words find themselves deficient. Music can partner with words, or even without words, to express levels and layers that have no direct translation. Music also brings a rich sonic tapestry to identities like religion, heritage, ethnicities, and regional areas. And some of the most interesting music happens when there is a fusion of these identities – when the spirit of one type of music melds with something totally distinct to create a new sound.

Music is also a tool to express and explore spirituality. It can help quiet our minds to have deeper conversations within. It can allow us to access deep information in our own brains when studying, meditating, or exercising. There is a spiritual connection also that we may feel with each other through music making. To harmonize with another person's voice is to join with them in a consonant and empathic way. Choirs aim to become 'one voice' through blending techniques and breathing in the same places. Orchestras and bands come together through the music on the page and the hands of their leader to make glorious and powerful musical assertions and stunning moments of sonic vulnerability.

To work with music and musicians is a spiritual practice. We all care so deeply about what we do and it is a great spiritual honor to the craft to carry this respect and acknowledgment of each other's humanity, love, and equality.

What Is Spirituality?

Spirituality is a grounding transcendent strength. Those who exhibit this in their top strengths anchor to a grander purpose and higher meaning in life. Spirituality is not tied to any one religion or practice, but rather a set of guiding beliefs that can include thoughts about belief systems, the universe, intentionality, connection, and goodness. Spiritual practices can also take many forms and many of them centers around music or use music as a tool to access higher powers or energies within. Think about any human ritual you have attended tied to a religious institution – most likely, the structure of it is tied strongly to music or musical cues. Even yoga, a moving meditation, and spiritual practice employs music or ambient sounds to deepen the connection. Either through God, nature, or personal spiritual practices, you find relationships and unity with yourself and the universe.

Spirituality connects us to a higher power, to each other, and aligns us with our own internal calling. It helps us make sense of the ups and downs and allows us to see life's challenges as lessons.

Spiritual people are often open-minded. They use many ways to express their goodness: accepting others, volunteering their time, helping those in need, and/or supporting worthy causes. Spirituality has the great capacity to bring groups of all sizes together. The importance of spirituality is magnified in difficult times. During COVID lockdowns, strangers in different cities and towns would join in spontaneous balcony music-making to find comfort, unity, and release. By connecting to a larger perspective, you become more resilient and can overcome hardships and challenges.

In today's contemporary world, there seems to be more spirituality and less formal religion. The chaos and pain of our contemporary world make living and surviving with some sense of peace and balance quite difficult. For many, spirituality has offered more freedom to personalize their own practices and create their own toolkit by experiencing many different approaches across religions, beliefs, and ancient and contemporary ideas. Spirituality exists in many creative forms including art, music, dance, voice, and even in the creation of electronic brain wave music designed to specifically soothe or stimulate parts of our brain.

Our modern world increased technology, bombardments of bad news, and the brain's hard-wired negativity bians has made us more susceptible to depression and other conflicts arising from societal, personal, and economic problems. Spiritual practices help us go beyond our daily unrest, ego, and struggles to attain some harmony and balance.

There are limitless options for spiritual expression. Backed by the sciences of happiness, character strengths, and whole being, we have realized that peace of mind is a proactive series of choices and behaviors we choose every minute of each day. One example of this is how meditation and mindfulness have become mainstream popular. Research has proven its value in extending lifespan, staying healthier, and having more creativity and success (Srivastava, 2017). Spirituality brings positivity into our lives, and we find our natural place in the universe whether in nature, chanting, or Sufi dancing (Brown, 2021)!

The advantages of spirituality are many. All means of expressing or experiencing spirituality provide a deep sense of grounding. It elevates optimism by providing a sense of purpose in life and adds to overall well-being. Spirituality is linked with many character strengths: humility, forgiveness, kindness, hope, love, zest, and gratitude. Spirituality also impacts behaviors activating compassion, altruism, philanthropy, and volunteerism. Spirituality promotes close family and friendship bonds and helps people cope with physical or emotional pain and other life stressors. It builds a warm sense of community.

Spirituality and Your Brain

It is proposed that spiritual practices have many beneficial effects on both physical and mental health. And yet, the exact neurobiology of spirituality remains largely unstudied. Imaging techniques have shown that gray matter increases after only eight weeks of meditation (*Change your mind: Meditation benefits for the brain*, 2022). While not conclusive, studies point to the involvement of the prefrontal and parietal cortices. As a result of meditative practices, there is a change in all the major neurotransmitter systems, resulting in decreased anxiety and depression, along with other health benefits (Krishnakumar et al., 2015).

Spirituality has been shown to shift the brain's focus to feelings of connection and community over feelings of stress (Gold, 2020). This occurs in the parietal cortex, an area of the brain responsible for attention processing and an awareness of self and others. Spirituality has long been used in healing, coping, and recovery. People turn to their spiritual beliefs in times of trauma and sadness. Studies show that people who lean on their spirituality show less anxiety, depression, and even suicide (Roberts, 2022).

The importance of spirituality has also revealed itself among populations of terminally ill patients (Chen et al., 2018). Incorporating an element of spirituality helped patients cope and become more resilient.

Affirmation

"I see and find meaning in my everyday life – both through my relationships and experiences."

Quick Tip

Look for the sacredness in your relationships and in every moment.

Musical Examples of Spirituality

- Working as a church musician
- Using songwriting to process a difficult event
- Thanking God at an award ceremony
- Starting practice with a prayer
- Getting lost in the Shavasana music at the end of a yoga class
- Dedicating a composition to someone
- Feeling driven or compelled to make music from an inner source or from God
- Losing oneself in an EDM (electronic dance music) song
- Finding one's purpose in playing their violin expressively
- Crying at the way music heightened a movie scene

Discussion Questions/Self-Reflections

Think about your close relationships with people, practices, or even music. Write about how those are sacred to you.

_____
_____
_____
_____
_____
_____

Treble and Bass

Treble • Overuse of Spirituality
FANATICISM: excessive and unquestioning enthusiasm or zest for something, such as a religion, political stance, band, or cause.

- Talent agent losing perspective about a band they love and represent.
- Choir director forcing religious-toned songs in a non-denominational setting.

Bass • Underuse of Spirituality
ANOMIE: an instability resulting from a breakdown of standards and values or from a lack of purpose or ideals.

- Business manager stealing money from their artist.
- Musicologist plagiarizing a lesser-known colleague's research.

Lights, Camera, Action!

- Create a playlist of songs that bring you closer to your own spiritual practice.
- Involve music in your yoga or meditation practice to help you go deeper inward.
- Feel the specialness of the present moment right now.

- Find meaning in a past hardship that has made you stronger today.
- Create your own definition for 'spirituality' in tandem or apart from religion.
- Ask someone of another religious faith about their beliefs.

Spirituality VIP
Yossele Rosenblatt • Worship Musician • Rabbi

Ukranian-born Yossele Rosenblatt (1882–1933) became one of the most famous voices in Jewish cantorial music; a rock-star rabbi. His voice was said to ring true to the words he was singing and people gathered to hear his heartfelt songs. His spiritual calling led him to a life of recording and touring around the United States and across Europe to full audiences. His vocal performances and compositions were said to represent the reality of prayer – not just the praise but also the struggle (Jaffe, 2010).

References

Brown, H. (2021, May 25). *Sufi whirling: These dancers can spin before they can talk.* euronews. Retrieved January 29, 2023, from https://www.euronews.com/travel/2021/05/15/sufi-whirling-meet-the-dancers-who-learn-to-spin-before-they-can-talk

Ask the Scientists (2022, July 20). *Change your mind: Meditation benefits for the brain.* Retrieved January 29, 2023, from https://askthescientists.com/brain-meditation/

Chen, J., Lin, Y., Yan, J., Wu, Y., & Hu, R. (2018). The effects of spiritual care on quality of life and spiritual well-being among patients with terminal illness: A systematic review. *Palliative Medicine, 32*(7), 1167–1179. 10.1177/0269216318772267

Gold, M. (2020, August 3). *How does spirituality change the brain?* Council on Recovery. Retrieved January 29, 2023, from https://www.councilonrecovery.org/how-does-spirituality-change-the-brain/

Jaffe, I. (2010, September 6). *Yossele Rosenblatt: The cantor with the heavenly voice.* NPR. Retrieved January 29, 2023, from https://www.npr.org/templates/story/story.php?storyId=129635330

Krishnakumar, D., Hamblin, M. R., & Lakshmanan, S. (2015). Meditation and yoga can modulate brain mechanisms that affect behavior and anxiety – A modern scientific perspective. *Ancient Science, 2*(1), 13. 10.14259/as.v2i1.171

Roberts, N. F. (2022, October 12). *Science says: Religion is good for your health.* Forbes. Retrieved January 29, 2023, from https://www.forbes.com/sites/nicolefisher/2019/03/29/science-says-religion-is-good-for-your-health/?sh=489df4673a12

Srivastava, P. (2017, March 10). *Importance of spirituality in the contemporary world – New Delhi Times – India only international newspaper.* New Delhi Times. Retrieved January 29, 2023, from https://www.newdelhitimes.com/importance-of-spirituality-in-the-contemporary-world123/

Gratitude

Musical Prelude to Gratitude

Gratitude is a centering strength. Acknowledging and giving thanks for the many gifts we are given in music – our talent, collaborators, bandmates, teachers, luck, and opportunities – powerfully amplifies your grace and attracts more of the good. You might notice that when we focus our attention on what we are thankful for, we tend to notice even more marvelous things about our daily happenings. Exercising gratitude helps us refocus our lenses to see the blessings and opportunities that surround us. Expressing thanks for the everyday bits of good fortune primes you to find more good to be thankful for. It's a multiplier that boosts the signal of everything wonderful in your life. Yes, it is true that there are many negative things that surround us, but the power of positivity, especially in music, is a superpower. The singer who loses their high register to a cold may be grateful that they still have their lower range and choose to transpose songs to explore tones there. They might come away from this experience with a newfound gratitude for their perseverance, new vocal palettes to explore, and a sense of resiliency. Gratitude helps play over all the noise so you can express yourself from a grounded, truer place.

Gratitude can be used as a tool to shift your focus. It is especially potent in combating stage fright and anxiety. When faced with stultifying fear, try counting your blessings. Concentrate on the fact that you are one of the lucky people who got to study their instrument. There are many parts of the world where survival is the mainstay of existence. Music is a luxury that you were able to partake in. Then think about how you up-leveled your skill set to the degree where you were able to make it part of your livelihood. And ultimately, you are lucky enough to get an opportunity to perform. Finally, you can be grateful that you get to share your love of what you do with an audience of others who are excited to hear you. This gratitude check-in can change your perspective quickly and thus, your fear factor may lessen. And, of course, gratitude is quietly present in the more mundane moments. It's important to stay grateful for the background privileges of your daily life – your health, family, pets, friends, job, home, city, etc. You always can find something you are grateful for.

Often gratitude goes undercover. Sometimes the lessons that are hardest or most painful to learn become our greatest moments of gratitude in disguise. They may feel excruciating at the time but end up being the most important growth plates which develop us further. This happens all the time in the field of music, with the many rejections and illusions that abound. Sometimes losing one job or deal turns out the be the greatest blessing because it recalibrates us onto another path that we are better suited for. Sometimes our subconscious sabotages us in roles we really don't fully feel or get behind. Think of musicians who failed in one genre and thrived in another that felt more aligned with them. Showing gratitude for both the good and the bad things that happen adds perspective and balance. Gratitude imbues faith in the process and allows you to trust the twists and turns of the musical journey.

A grateful spirit attracts bounty and abundance. Being grateful for what you have is not an admission that you are satisfied with everything the way it is now. It is rather an acknowledgment that you respect the gifts you do have right now and are in alignment with the energy of receiving more of the same. That warm spirit opens you up to receive gratefully. Be grateful for the artists you are producing today and let them feel your thanks. This paves the way to bring more and bigger artists your way. Remember, showing gratitude is free – it is the one thing you have that you can give away and it never will deplete you.

What Is Gratitude?

Gratitude is the quality of being thankful, giving kindness, and showing appreciation. It is the queen of all virtues and the heavy hitter of all the character strengths! Gratitude also looks for the good in everything and everyone. It is always present despite the loud chatter and blasting music of our negativity bias. There is more research on gratitude than most of the other character strengths and topics in positive psychology. People understand gratitude intuitively and culturally and it is a fast and effective way to boost happiness and well-being.

Gratitude opens our eyes to the gifts around us – both large and minute. By focusing on these, we view our lives as rich and abundant. It's easy to become numb to the everyday regular sources of goodness in our lives and that is why surprises tend to bring stronger feelings of gratitude.

Gratitude has two parts: an affirmation of goodness and an appreciation of where that goodness comes from. A grateful person acknowledges the struggles of life and still holds the perspective that, in its totality, life is good. They affirm that this goodness is always there as a trusted canvas under the sometimes chaotic-looking painting of life. This goodness is often independent of our actions. It comes from other people, higher powers, or spiritual beliefs that help us to feel connected and achieve great things.

Many studies have looked at the effects of gratitude journals and practices on their ability to improve our health, happiness, and relationships (Komase et al., 2021). Results are consistent and overwhelming that even after just three weeks of practice on a regular basis, gratitude practices improve well-being, relationships, and physical and mental health.

In 2015, University of California Davis professor and leading expert on the science of gratitude Dr. Robert Emmons and University of Miami professor Dr. Michael McCullough published a study that looked at the physical outcomes of practicing gratitude (Emmons & McCullough, 2003). Subjects were divided into three groups. The first group kept a daily gratitude journal, the second kept a daily irritation journal, and the third neutrally journaled with no directive. After ten weeks, the first group reported feeling the most positive and optimistic about life and were more physically healthy and active than the other two groups.

Some of the physical benefits of gratitude are fewer aches and pains, stronger immune systems, lower blood pressure, sleeping longer and feeling more rested, increased exercise, and overall better health (*Gratitude is good medicine*, 2015). Psychological benefits are an overall higher level of positive emotions such as joy, pleasure, optimism, happiness, and feeling more alert and alive.

Gratitude is a prosocial positive emotion and character strength that elicits more generous, helpful, and compassionate behavior. It explores how we are supported and affirmed by others. It is a relationship-strengthening practice resulting in people feeling less lonely and isolated. Doctors John and Julie Gottman found that healthy couples express their gratitude on a regular basis. One simple Gottman Method for a gratitude practice is to have partners say 100 positive or kind things to each other in one day (*The Gottman method*, 2022). When you see and acknowledge all the good, you grow as an individual and as a couple – and it feels great! Or as Dr. Tal Ben Sha-har often says, "When you appreciate the good, the good appreciates" (*Appreciate the good with Dr. Tal Ben-Shahar*, 2018).

Gratitude and Your Brain

Gratitude rewires your brain to be happier by causing synchronized activity in multiple brain regions. It lights up the hypothalamus and parts of the brain's reward pathways.

The networks that fire during pleasure and socialization also light up with gratitude.

Experiencing gratitude boosts the neurotransmitter serotonin which then activates the brain stem to produce dopamine – the brain's pleasure hormone (*How to practice gratitude*, 2022).

Neuroscientists uncovered a remarkable distinction in expressing gratitude. It has been shown that those who engage their emotions when expressing thanks and take time to really "feel it" experience more benefits than those who just say thank-yous to be polite. The regular practice of gratitude reaps true health and happiness benefits for those who incorporate it like a daily vitamin (*Gratitude literally rewires your brain to be happier*, 2022). And perhaps most empowering, Dr. Rick Hanson teaches us that for our brain to "let it in" we must keep experiencing the thoughts, feelings, and images for at least one to two minutes (Hanson, 2021).

Affirmation

"Gratitude is a gift that must be opened. I appreciate and express my thanks every day."

Quick Tip

Gratitude is so good for you! It's a big-bang practice with numerous physical, psychological, and social benefits.

Musical Examples of Gratitude

- Giving gifts to teachers before they leave for the holidays
- Expressing heartfelt thanks after an orchestral recording session

- Thanking marching band members for a taxing practice
- Acknowledging your management team while accepting an award
- Thanking your parents' financial sacrifices at your senior recital
- Appreciation of one's health and ability to perform
- Staying grounded during a meteoric rise to fame
- Giving a free concert in the neighborhood in which you grew up

Discussion Questions/Self-Reflections

Keep a gratitude journal. Briefly, write five things you experienced in the last week for which you are grateful. Research shows you can get the most benefits from this experience by doing this at least three times a week. Also, think about who contributed to your gratitude experience and how they acted. Think of each experience on your list as a gift. Do this with deliberate slowness, call up the visual memory, and recall your thoughts and feelings at that moment. Deeply feel your thankfulness as you write.

_____
_____
_____
_____
_____
_____

Treble and Bass

Treble • Overuse of Gratitude
INGRATIATION: an overuse of flattery to garner likeability.

- Manager too worried about upsetting their client that they don't advise them against a decision they shouldn't make.
- Teacher too concerned about being a friend that they forget to be a mentor.

Bass • Underuse of Gratitude
RUGGED INDIVIDUALISM: focus on self-reliance, often a blind spot where one does not see the contributions of others to their success.

- Orchestral conductor who takes all the credit for the work of the orchestra.
- Playlist curator who misses new bands because they refuse to research new trends and ignore submissions in genres with which they are not familiar or already fluent.

Lights, Camera, Action!
- Send thank you messages to everyone who was part of a recent successful project (album launch, live show, recording session).

- Wear a piece of clothing (shirt, scarf, accessory) or jewelry that reminds you of the people who helped you get to where you are now (like the character strengths bracelets!). Objects, symbols, and accessories are great primers and prompts to help our brain integrate and support our practices.
- Find someone in your life who is not typically recognized and deserves to be. Tell them thanks and name specifically what you are thanking them for.
- As a surprise, spontaneously put a thank-you note on someone's kitchen counter, notebook, purse, or desk.
- Send a spontaneous gratitude text or email every day for one week!
- Keep a gratitude journal every day for a week and notice your perspective shift.

Gratitude VIP

Pharrell Williams
Producer • Songwriter • Performer • Entrepreneur
• Philanthropist • Designer

It is hard to find an interview or interaction with multi-hyphenate, award-winning producer/songwriter Pharrell Williams that doesn't begin with a "thank-you" or a very specific declaration of gratitude to the interviewee or fellow producer. Williams' record-breaking hit "Happy" unleashed a wave of joy and resulting gratitude that swept the planet. In an emotional interview with Oprah, Williams stated, "I appreciate the fact that people have believed in me for so long that I can make it to the point to feel that" (YouTube, 2014), referring to the joy and gratitude he felt while watching a video of people all over the world expressing happiness to his song. Amongst the many businesses and endeavors in his empire, Williams added a Masterclass series on empathy, where he describes how he uses it to connect to others with music (*Waking up to empathy*, n.d.). Evidenced by his many charitable acts and ways in which he is constantly giving back to his hometown of Virginia Beach, Williams keeps gratitude at the forefront of his actions and deeds (Bhattacharji, 2021).

References

Bhattacharji, A. (2021, December 8). *Is Pharrell Williams the world's best neighbor?* Town & Country. Retrieved January 29, 2023, from https://www.townandcountrymag.com/society/money-and-power/a36489849/pharrell-williams-interview-black-ambition/

Emmons, R. A., & McCullough, M. E. (2003). Counting blessings versus burdens: An experimental investigation of gratitude and subjective well-being in daily life. *Journal of Personality and Social Psychology, 84*(2), 377–389. 10.1037/0022-3514.84.2.377

UC Davis Health (2015, November 25). *Gratitude is good medicine.* Retrieved January 29, 2023, from https://health.ucdavis.edu/medicalcenter/features/2015-2016/11/20151125_gratitude.html

NeuroHealth Associates (2022, May 6). *Gratitude literally rewires your brain to be happier.* Retrieved January 29, 2023, from https://nhahealth.com/neuroscience-reveals-gratitude-literally-rewires-your-brain-to-be-happier/

Hanson, R. (2021, December 15). *Take in the good*. Dr. Rick Hanson. Retrieved January 29, 2023, from https://www.rickhanson.net/take-in-the-good/

Mindful. (2022, October 4). *How to practice gratitude*. Retrieved January 30, 2023, from https://www.mindful.org/an-introduction-to-mindful-gratitude/

Komase, Y., Watanabe, K., Hori, D., Nozawa, K., Hidaka, Y., Iida, M., Imamura, K., & Kawakami, N. (2021). Effects of gratitude intervention on Mental Health and well-being among workers: A systematic review. *Journal of Occupational Health*, 63(1). 10.1002/1348-9585.12290

The Gottman Institute (2022, August 4). *The Gottman method*. Retrieved January 29, 2023, from https://www.gottman.com/about/the-gottman-method/

MasterClass (n.d.). *Waking up to empathy*. Retrieved January 29, 2023, from https://www.masterclass.com/classes/the-power-of-empathy-with-pharrell-williams-and-noted-co-instructors/chapters/waking-up-to-empathy

YouTube (2014). *"Happy" makes Pharrell Williams cry*. YouTube. Retrieved January 29, 2023, from https://www.youtube.com/watch?v=IYFKnXu623s.

YouTube (2018). *Appreciate the good with Dr. Tal Ben-Shahar*. YouTube. Retrieved January 29, 2023, from https://www.youtube.com/watch?v=jYRJEi-9tiE.

THE HUMANITY STRENGTHS

Love

Musical Prelude to Love

How many musicians and music professionals can point to love as their raison d'être? We fall in love with music at an early age. If we are lucky enough, we are encouraged and nurtured to pursue this love through formal study, immersive experiences, and perhaps later, majoring in music at a college or university. Others take their talent directly to the business-facing side of music and waste no time, working with other musicians and creating things to put out in the world.

Love and true connection are valiant goals, whether within your artistry, relationships, or career. When you really love what you do, you want to pour yourself into it. There is a sense of joy and flow in your actions and the way you do things. Sometimes that takes clock time to develop. When first starting out, teachers encounter many difficulties and challenges. Learning the ecosystem and politics of a school, how to appropriately allot time, lesson planning, dealing with behavioral issues, parents, and even memorizing names are huge learning curves that every beginning teacher must slowly roll through. At these stages, the teaching itself might take a backseat to these other factors and mastering the proper execution of pressing classroom management matters. Though the love is there, the act of teaching it is not yet in flow and doesn't have automaticity. It's a real challenge at first. As they develop their style and gain the experience that comes with time and encountering scenarios, this begins to shift, and they can access their love of the craft once again.

Music also acts as a megaphone for love; magnifying the way we express it. Whether vicariously by playing/listening or directly through our own hands and voices, music provides a platform for all love's "feels." By volume, most songs deal with some aspect of love – finding it, keeping it, losing it, wanting it, or missing it. These universal experiences bring so many people to grieve, celebrate, and pray to the altar of these songs. The pain of emotional loss finds solace on the musical page and the joy of new love flies off it. And there is nothing like the literal playlist sharing that occurs when you fall in love with someone. As if bringing them closer to the music you love somehow is a shortcut to your soul itself.

There are industries and businesses that you don't have to engage with love. You may get into them because you have an interest, proclivity, or skill set, or know that they will provide you with the lifestyle you want. Music is not one of those. You must have a love for music to be involved with it at the professional level. And no matter what job you have within music, you can trace it back to some love affair of your formative years. Creative spaces burn brightly with passion and people who will push you far past the places you thought you would end up.

Connect often to things you love – those things keep you as alive as you can be. And renew your vows with music as often as you need to. Let it remind you of your once and always true love.

What Is Love?

Love is one of those words that has many meanings and layers of complexity. There are probably more poems, books, plays, and music about love than any other topic. According to ancient Greek culture, there are many types of love (Krznaric, 2020). They distinguished between passionate sexual love (eros), the deep love for your friends or fellow soldiers (philia), playful and flirty love in the early stages of courting (ludus), a universal love for mankind (agape), long-term love that has withstood sacrifice (pragma), and love of one-self (philautia). For such a complicated set of emotions, it makes sense to have these many flavors and qualities expressing differing types.

Similarly to the Greeks, author and pastor Gary Chapman wrote about behaviors he observed in couples he counseled and coined the "Five Love Languages" (Chapman & Summers, 2010). These are words of affirmation, quality time, physical touch, acts of service and receiving gifts. There are quizzes online you can take to discover your own language. Think about this conceptually – we speak in one love language and our bandmates or colleagues may be speaking in another. As an act of care, take time to glean the 'love languages' of those around you and then act in their love language to connect more closely. Surprise someone you love with a small, unexpected gift; flowers, a treat, or food you know they like. Speak your love in words and tone. Write them a wonderful letter or text. You can even comment on a strength you appreciate about them – and be specific.

In a healthy relationship, love is reciprocal – there is a balance between giving and receiving. Also in love, healthy interdependency and autonomy co-exist in balance. Many musicians will cite love as the number one reason for getting into the profession. You may even hear people speak about 'falling in love with their instrument' the moment they either heard someone play it or discovered it for themselves. Creatives often describe their relationship with music and the music business side as a love/hate relationship. Perhaps, we could describe the love one feels for music as a long-term commitment filled with ups and downs and sacrifices (time, money, energy, and effort) akin to the "pragma" kind of love.

Ryan Neimiec discusses love as a fundamental characteristic of positive relationships and social functioning (Niemiec, 2018). Those who are high in the strength of love have higher self-esteem, less susceptibility toward depression, and a better capacity to cope with stress.

Love and Your Brain

Science has confirmed that love can actually change our brains, reduce stress and pain, and can even help us live longer (*The health benefits of doing what you love*, 2019). Falling in love has a similar chemical effect to the euphoria created by substances such as cocaine (Springer, 2012), and falling in love replicates the body's drug addiction state (Tigar, 2016). Both drug usage and love trigger the release of a trilogy of our happy hormones: dopamine, oxytocin, and vasopressin as well as adrenaline. No wonder people look and ask, are you in love? You look so amazing, happy, and high on life!

High levels of dopamine activate the reward circuit in the brain making love a pleasurable experience like the euphoria of alcohol or cocaine (Edwards, 2015).

Oxytocin, or the "cuddle hormone," is associated with social connection and it is released during sex and other physically connective acts such as hugging. Vasopressin has a similar effect and is related to romantic love. Women tend to be more sensitive to oxytocin and men to vasopressin (Dockrill, 2016).

Dopamine fires up relationships and maintains them. Oxytocin receptors in the brain region mediate reward and motivation. Research has shown that people who have a greater amount of oxytocin receptors in the brain, have a higher likelihood of forming lifelong pair bonds (Bosch & Young, 2017).

Being in a loving relationship impacts us positively both emotionally and physically. Love fuels health and longevity and advances empathy, forgiveness, acceptance, and tolerance. Love is amongst the top five character strengths associated with life satisfaction.

Affirmation

"I give and receive love freely."
"I show my love by making the needs of the other person as important as my own."
"I listen deeply with warmth and curiosity."

Quick Tip

Love takes time to grow. Start by making the needs of the other person as important as your own.

Musical Examples of Love

- Passion for your favorite artist's new release
- Songwriting about love and relationships
- Loyalty to a choir or a cappella group
- Forming bonds with your fellow musicians
- Researching a subject or composer deeply and with care to detail
- Taking on free gigs just for the love of playing or for charity
- Passing on an opportunity of your own to help feature someone you care about
- Working in A&R and living and breathing new artists/music
- Authoring a music blog or critique column
- Teaching music

Discussion Questions/Self-Reflections

Are you more comfortable giving or receiving love? Reflect on what makes this so. Sometimes we are better at giving than receiving, or vice versa. How could you balance giving and receiving more?

_____
_____
_____
_____
_____
_____

Treble and Bass

Treble • Overuse of Love

EMOTIONAL PROMISCUITY: you use the word *love* freely, with everyone to the point that it has little meaning, not very discriminating in your choices.

- An agent who pretends to care very deeply about each client and uses parabolic praise and over-promises with little result.
- A teacher who lacks appropriate boundaries and allows their moods and personal life to bubble over into the classroom experience.

Bass • Underuse of Love

EMOTIONAL ISOLATION: detachment, the stance that people and relationships are messy and not worth it.

- A concert pianist who spends all their time alone in a practice room and can't connect with their audience or other musicians.

- A producer who spends the bulk of their time on their computer but not interfacing or interacting with the people in the room.

Lights, Camera, Action!

- Sing a song just because you love it.
- Help your music therapy clients build a repertoire that brings them joy.
- Spontaneously do something nice for the people you work with like restock a snack jar.
- Take time to write an encouraging or positive comment on a YouTube video you like.
- Send the artists on your roster an encouraging message "just because."

Love VIP

Jathan Wilson
Creative Direction • Marketing

Jathan Wilson, senior manager of Creative Strategy and Marketing at Disney Music Group, fell in love with one of his latest projects collaborating with ESPN. *The Undefeated* is a series focusing on the symbiotic relationship between Black athletes and music for change in their communities (Nwulu, 2021). He speaks with such care and enthusiasm for this meaningful endeavor and describes the whole process in a way people explain the flow state, saying that the whole execution of the musical side only took five weeks, which is light-ning fast for these types of projects. Wilson talks about the importance of being a Black man in the industry and working on projects that uplift his community. He also speaks highly of the seamless collaboration of the entire team that helped bring this to fruition. When love is guiding a project, and all parties are equally invested, the results are powerful and transformative.

References

Bosch, O. J., & Young, L. J. (2017). Oxytocin and social relationships: From attach-ment to Bond Disruption. *Behavioral Pharmacology of Neuropeptides: Oxytocin*, 97–117. 10.1007/7854_2017_10

Chapman, G. D., & Summers, A. (2010). *The Five love languages: How to express heartfelt commitment to your mate*. LifeWay Press.

Dockrill, P. (2016, May 20). *What is love, anyway? here's the science*. ScienceAlert. Retrieved January 29, 2023, from https://www.sciencealert.com/what-is-love-anyway-here-s-the-science

Edwards, S. (2015). *Love and the brain*. Harvard Medical School. Retrieved January 29, 2023, from https://hms.harvard.edu/news-events/publications-archive/brain/love-brain

Krznaric, R. (2020, December 7). *The ancient greeks' 6 words for Love (and why knowing them can change your life)*. YES! Magazine. Retrieved January 29, 2023, from https://www.yesmagazine.org/health-happiness/2013/12/28/the-ancient-greeks-6-words-for-love-and-why-knowing-them-can-change-your-life?gclid=CjwKCAjwlqOXBhBqEiwA-hhitJwfLJj0FZ30E4Iqup0-Z_Jwmo6f9G54RRw7w0oLD27-_Qv5YG5tuRoC9woQAvD_BwE

Niemiec, R. M. (2018). *Character strengths interventions a field guide for Practitioners.* Hogrefe.

Nwulu, M. (2021, February 26). *Sunday's 'A room of our own' special on ESPN powered by new covers of classic songs.* ESPN Front Row. Retrieved January 29, 2023, from https://www.espnfrontrow.com/2021/02/a-room-of-our-own-special-on-espn-powered-by-new-covers-of-classic-songs/

Springer, S. H. (2012, August 4). *Falling in love is like smoking crack cocaine.* Psychology Today. Retrieved January 29, 2023, from https://www.psychologytoday.com/us/blog/the-joint-adventures-well-educated-couples/201208/falling-in-love-is-smoking-crack-cocaine

Thrive (2019, August 22). *The health benefits of doing what you love.* Retrieved January 29, 2023, from https://thrive.kaiserpermanente.org/thrive-together/live-well/health-benefits-of-doing-what-you-love

Tigar, L. (2016, February 12). *How your body reacts when you fall in Love.* CNN. Retrieved January 29, 2023, from https://www.cnn.com/2016/02/12/health/your-body-on-love#:~:text=Being%20in%20love%20is%20like%20a%20drug%20addiction&text=Researchers%20concluded%20that%20falling%20in,oxytocin%2C%20adrenaline%2C%20and%20vasopressin

Kindness

Musical Prelude to Kindness

Making and sharing music is something special that comes from a very alive place in us. It takes bravery to devote your life to this pursuit which requires a multiplicity of roles and jobs. It is not a path paved with certainty, but rather with tons of prospecting, work, hope, and risks. Kindness is a value that provides comfort and humanity during this upward climb. A cup of kindness goes a long way on the steep and winding trails to our musical dreams. We are often hardest on ourselves, so it is not only a cup to share but one to sip from. We need to hold space to be kind to our bodies, minds, and spirits as we endeavor our musical aspirations.

Though music starts as a painting of our own souls, it becomes a canvas that others will experience and filter through their own emotional painting. That's a doubly vulnerable state, this relationship between the creator, the creation, and the receiver and thus the re-creator of the work of art or music. To build respect and trust in this sacred space, it is imperative to keep kindness at the forefront of one's mind and treat your listeners, audiences, and students, with gentleness.

Kindness is also free. There is no shortage of negative feedback in the music industry. The critical voice, from within and externally follows each project with comparisons, judgments, and opinions that can leech the joy out of the mix. Bringing a kind voice to the most unfavorable feedback makes a very real difference in the way a person will receive it and respond. You can easily shut down an artist by criticizing their work. So much of music comes from a deep place inside an artist, it intermingles with their identity and self-worth. They may experience the criticism as very personal and pejorative. Putting space between the work and the artist is very important and taking the time to put yourself in the other person's shoes can show you how you might want to hear and therefore deliver bad news. Kindness in this way needs social intelligence to inform and support it.

Kindness is also contagious! It feels so great to make other people feel wonderful, it almost seems like a selfish act. There are many ways to exhibit kindness in the music profession. You can give your advice freely to young people studying in your field. You can demonstrate by showing kindness to up-and-coming performers and give them performance opportunities they might not otherwise encounter. You can donate your skill set to tutor or teach youth who don't have funding, support, or access to the arts or instruments. You can also help connect a struggling musician with a job connection. The ways are endless and all equally uplifting and joyful.

So yes, we may tune into the competition show to hear what the blunt judge will say, but at the end of the day, true acts of kindness are more long-lasting and impacting than negative ones. You have a heavy hand in choosing the ways people will remember you. A legacy of kindness is a lovely one to leave behind.

What Is Kindness?

Kindness is a combination of nurturing, generosity, empathy, and altruism. Kind people make the world a sweeter and safer place. They easily and freely help or do favors, even without knowing the other party well or at all.

Though music is a beloved passion for many, the journey into working in music is quite long, intense, and deceptively arduous with many distractions, pitfalls, and important lessons along the way. We can often think back to someone whose kindness to us in our formative years made a big difference along that path.

Kindness is hard-wired into our psyche with the message that helping others and working together is the right thing to do. It is also spiritually satisfying for some. The development of kindness is important on an individual as well as a societal level. Studies have shown that kindness is contagious and create a "cascade" effect (Gregoire, 2015). When you witness kindness in action, it engenders feelings of connection and warmth that incline others to do a kind act as well. Kindness is contagious, creating a chain reaction of good deeds. Why is this so? Just as working out muscles makes them stronger; the consistent activation of the same brain circuits makes them stronger and easier to access.

With consistent kindness, we come to genuinely care for others more and find it easier to access happiness. In turn, extracting happiness from the everyday backdrop of events and circumstances in our lives becomes more natural. Smiles become easier and more natural too. And it's all good for our health. It has been shown to make one healthier and even slow the aging process (Le Nguyen et al., 2019).

Kindness is seen to be an important element in society, especially in public policymaking (*Why kindness matters in public policy*, n.d.). Movements like "random acts of kindness" were created to connect people through deliberate acts of care and generosity by collecting these acts and sharing them on media or websites. When we watch or see acts of kindness, we are more inclined to want to do them, too (*Welcome to Randomactsofkindness.org*, n.d.)!

Kindness studies ask people to commit to doing a prescribed number of acts of kindness over a period – a week, month, or longer. Happiness levels are measured before and after the study and then compared to people not doing kind acts. While there are many versions of these studies, they all agree that being kind increases our happiness (Stoerkel, 2022). Above all, kindness feels good and improves relationships by creating emotional warmth.

Kindness and Your Brain

Kindness impacts our brains in so many positive ways. Dr. David Hamilton explains that we have "kindness genes" and they are nearly 500 million years old in human genomes (Hamilton, 2021). Kindness has played a massive role in our survival. We are genetically drawn to help others.

The brain networks involved in developing kindness are complex. Oxytocin (the main kindness hormone released in love, drugs, and bonding) works to dim down activity in the amygdala. This amygdala is involved with anxiety, depression, stress, worry, and fear. With kindness on board as one of your strengths, these stresses don't land as hard. Oxytocin also lowers blood pressure and increases heart health (Owens, 2021). Kindness chemicals and hormones impact your mood and produce a sense of well-being. Also, the benefits of kindness are not only experienced by doing these acts but also by simply witnessing them. Studies show people of all ages respond with greater well-being, happiness, and life satisfaction (*The science behind kindness and how IT benefits your health*, 2020).

Research in neuroscience has shown that kindness can be trained and can change the structure and function of our brains. This malleability of the brain is known as neuroplasticity (Josipovic, *The neuroscience of kindness*).

The benefits of kindness are far-reaching. Studies show that people who carry out acts of kindness experience greater well-being, happiness, and life satisfaction. When you are kind, your brain's pleasure and reward centers light up – this is called "helper's high" (Baraz & Alexander, 2010). This is true whether you are the giver or recipient of the good deed (Angier, 2009).

Volunteering is a great way to put kindness into action. People who volunteer experience less aches and pains. People 55 years and older who regularly volunteer have a 44% lower likelihood of dying early. Surprisingly, volunteering has a stronger effect than going to church or exercising four times a week (Carter, 2010).

Affirmation

"I strive to be kind to everyone I meet."
"I aim to only speak kind words to and about others."

Quick Tip

Kindness is almost always the right choice, and it feels great!

Musical Examples of Kindness

- Working patiently with children in the classroom
- Teaching special education classes
- Being a music therapist
- Playing for charity events
- Volunteering to play at a nursing home
- Giving compliments to your fellow musicians
- Donating your musical services
- Seeking out underrepresented composers or demographics
- Understanding your students' hardships
- Giving musical equipment away
- Subbing in for a teacher's class who called in sick
- Filling in for a gig or rehearsal for a friend

Discussion Questions/Self-Reflections

Think of a kind act that was done for you. Who did it and how did it make you feel? What are three things you can do today to practice kindness?

_____
_____
_____
_____
_____
_____

Treble and Bass

Treble • Overuse of Kindness

INTRUSIVENESS: overly caring or inserting yourselves into the private matters or lives of others.

- Private lesson teacher who insists on knowing the emotional details of your personal life.
- Music librarian who unsolicited injects themselves into your research because they want to help.

Bass • Underuse of Kindness
INDIFFERENCE: uncaring, paying little attention to anything that doesn't directly benefit you.

- College professor who belittles "new" music without listening to it.
- Choir director who has students sing when they're not feeling well.

Lights, Camera, Action!

- Volunteer your services or talents to a worthy cause or population in need (hold a free concert at a nursing facility, raise money for the homeless though music).
- Donate music equipment or books to a school in need.
- Serve meals at a local shelter during the holidays (or any day).
- Get a few gift cards and give them to someone homeless you notice on your everyday route.
- Write out a few kind messages on cards and put them on parked car wind shields.
- Develop a meditation akin to the powerful Buddhists' Loving Kindness meditation. In this practice, we wish others, ourselves, and the world happiness, good health, and freedom from suffering.

Kindness VIP
Mary Rudenberg • Music Therapist

Pioneering music therapist Mary Rudenberg's life work reflects kindness through expressive therapies. She is most known for her work with children having cerebral palsy and other orthopedic and mental handicaps. She used her love of dance and the flute as a tool for the healing of others. In an interview, Mary was asked to describe a peak experience she had as a music therapist. She spoke fondly of a dance performance she worked on with patients for weeks. This included a blind girl for whom she had specially choreographed a part. She gained a lot of satisfaction from seeing how effective and meaningful the whole dance was for the participants and her loving kindness was at the front and center of this activity (Neugebauer, 2010). Mary was also an active volunteer, serving in her church choir, city committees, hospices, and pediatric associations. She is a wonderful example of kindness begetting more kindness.

References

Angier, N. (2009, November 23). *The biology behind the milk of human kindness*. The New York Times. Retrieved January 29, 2023, from https://www.nytimes.com/2009/11/24/science/24angier.html?_r=1&partner=rss&emc=rss

Apple Podcast (2020, October 7). *The science behind kindness and how IT benefits your health*. University Hospitals. Retrieved January 29, 2023, from https://www.uhhospitals.org/blog/articles/2020/10/the-science-behind-kindness

Baraz, J., & Alexander, S. (2010, February 1). *The helper's high.* Greater Good. Retrieved January 29, 2023, from https://greatergood.berkeley.edu/article/item/the_helpers_high

Carter, C. (2010, February 18). *What we get when we give.* Greater Good. Retrieved January 29, 2023, from https://greatergood.berkeley.edu/article/item/what_we_get_when_we_give

Gregoire, C. (2015, May 19). *Why kindness is contagious, according to science.* HuffPost. Retrieved January 29, 2023, from https://www.huffpost.com/entry/kindness-contagious-psych_n_7292862

Hamilton, D. (2021, July 26). *Wired for kindness.* David R Hamilton PHD. Retrieved January 29, 2023, from https://drdavidhamilton.com/wired-for-kindness/

Josipovic, Z. (n.d.). *The neuroscience of kindness.* UNESCO MGIEP. Retrieved January 29, 2023, from https://mgiep.unesco.org/article/the-neuroscience-of-kindness

Le Nguyen, K. D., Lin, J., Algoe, S. B., Brantley, M. M., Kim, S. L., Brantley, J., Salzberg, S., & Fredrickson, B. L. (2019). Loving-kindness meditation slows biological aging in novices: Evidence from a 12-week randomized controlled trial. *Psychoneuroendocrinology, 108,* 20–27. 10.1016/j.psyneuen.2019.05.020

Neugebauer, C. (2010). *Mary Rudenberg, music therapist Pioneer.* Voices. Retrieved January 29, 2023, from https://voices.no/index.php/voices/article/view/1856/1620

Owens, A. (2021, September 23). *Oxytocin: What it is, how it makes you feel & why it matters – psycom.* Retrieved January 29, 2023, from https://www.psycom.net/oxytocin

Stoerkel, E. (2022, September 10). *Can random acts of kindness increase wellbeing? (+ 22 ideas).* PositivePsychology.com. Retrieved January 29, 2023, from https://positivepsychology.com/random-acts-kindness/

The Random Acts of Kindness Foundation (n.d.). *Welcome to Randomactsofkindness.org.* Retrieved January 29, 2023, from https://www.randomactsofkindness.org/about-us

Mental Health Foundation (n.d.). *Why kindness matters in public policy.* Retrieved January 29, 2023, from https://www.mentalhealth.org.uk/our-work/policy-and-advocacy/why-kindness-matters-public-policy

Social Intelligence

Musical Prelude to Social Intelligence

Social intelligence might be the most important aspect of healthy human relationships. It is the all-access pass to knowing people and knowing yourself. We grow and build this by paying attention in situations, getting involved in various groups and entities, and dealing and working with a multitude of different people. There is nothing better than being in a partnership with someone who truly sees you, gets you, can anticipate your needs, and responds to them at face value without keeping score. To be on the giving end of this dyad, you need to have done your homework on the other person. Social intelligence is in part innate but can also improve with conscious work and practice.

Being able to read the room is of high value in the music industry. It allows you to make choices in the moment that will be more favorable for your intended outcomes. If you are in the middle of playing set of music that is not going as well as you planned, you can use your social intelligence to read the crowd, gauge the emotion of the room, and make a sharp left turn to play songs that will vibe with that.

Social intelligence is the wherewithal factor of establishing yourself in a community – be it a genre of music, an assembly of music technologists, a high school marching band, or the management team of a famous performer. Knowing how to spend and save your human capital depending on varying factors and changing personnel is key to having staying power in that circle. It is as important to know when not to show up for something as it is to show up. Having great social intelligence and discernment informs where you spend that effort.

A big part of social intelligence is the willingness to learn and grow from the experiences you have. Every performance is an opportunity to study how you are received. So is every release, social media post, interview, and interaction with varying parts of the business. And you don't have to be an expert at these things – this is where teamwork and trust come into play. As an artist, you want to assemble people around you who have your best interests in mind and can advise you when to push, pull back, make an introduction, show your music, and delay a release. There are musicians who hear but don't take the good advice from their teams. Perhaps they get lazy about promoting their next project to their fans, or they insist on over-playing their market, or they release something sub-par just to "get it out there." The lessons of social intelligence come in many ways and eventually, a self-reflective artist will examine their behaviors and impulses and weigh them against their results. They will have to learn to hear the unspoken and take cues from people's energy. Those who do not have great social intelligence would be wise to work with people who do and form a real trust and take notes. Otherwise, it makes getting to the next level that much tougher.

What Is Social Intelligence?

Social intelligence is the ability to understand and navigate interpersonal relationships. It involves a complex set of skills and understanding. It is the superpower that allows you to "read the room." Some of it is innate however it can also be developed as one learns from the successes and failures of their experiences. This strength is sometimes referred to as "street smarts" as it is not learned from a book but rather gained from social interactions. Becoming more aware of others' motivations and feelings can empower one to maneuver new situations. Musicians gain a "musical social intelligence" with each new gig, contract, opportunity, and performance they experience. There are notorious stories of young or inexperienced creators being taken advantage of because of their lack of gained social intelligence.

Social intelligence is a whole being strength with social, emotional, and health benefits (*What is Social Intelligence and why should I develop it?*, n.d.). The ability to navigate social settings helps one to build bonds and work well together (*6 ways to improve social skills and increase social intelligence*, n.d.). When relationships are more in flow, it is easier to deal with challenges and mitigate obstacles. Studies show people with high social intelligence are more optimistic and happier with their lives overall (Rezaei & Bahadori Khosroshahi, 2018).

Social intelligence is an interpersonal muscle-builder that helps you contain your emotions for the greater good. For some this comes easily, for others it takes a lot of practice and discipline. Socially intelligent people can be great communicators. A core of this is listening without interrupting and showing compassion and curiosity when the other person speaks. Then you think before speaking or acting, tune into your honesty and authenticity, and communicate as kindly and clearly as possible.

Social Intelligence and Your Brain

The frontal lobes of your brain control social intelligence, as well as other mature brain executive, functions such as planning, working memory, and impulse control (Sowell et al., 1999). Your brain continues to mature and develop through your twenties up until 30 years of age.

The frontal lobes are the area where social skills like motivation, personality, carrying out a process of several steps, social graces, and knowing the situationally appropriate behavior all reside. Research shows that people with stronger social connections, are more optimistic and positive, tend to suffer less from mental health problems, experience less stress, and consequently have better physical health (Rezaei & Mousanezhad Jeddi, 2018).

Affirmation

"I am tuned into my feelings and thoughts about myself and others."
"I listen and observe first before responding, so that I can do so in an appropriate manner."

Quick Tip

Take a walk in another person's shoes before deciding how you think or feel about them and their actions.

Musical Examples of Social Intelligence

- Reading a crowd and making a different, on-the-spot performance choice
- Captivating a classroom full of different learners and varying ability levels
- Delivering an intimate or sensitive lyric
- Fundraising by understanding the context of many social settings
- Dealing sensitively with management, the bride, or any other contractors
- Having close bonds with your students
- Listening to and learning from others
- Understanding how to best delegate roles in a production, event, or crisis
- Volunteering to speak to difficult conductors to voice group concerns
- Practicing entertainment law successfully
- Running your department without ego and embracing the philosophy that "other people matter"

Discussion Questions/Self-Reflections

Tune into someone else you are working or creating with currently. Take a few minutes to mentally walk in their shoes. What do you know about them? What's going on in their lives? What stressors might they be going through? Write down anything you can think of about them. What new information did this reveal? Does this create any shifts in you, especially ones that lead you to more empathy?

_____
_____
_____
_____
_____
_____

Treble and Bass

Treble • Overuse of Social Intelligence

OVER-ANALYSIS: thinking too much, overly cautious or inhibited.

- Recording engineer who can never release a track because it's not "perfect."
- Violinist who won't allow themselves to let go onstage for fear of making a mistake.

Bass • Underuse of Social Intelligence

CLUELESSNESS: naïve, thoughtless, emotionally insensitive, obtuse.

- Artist who says something in an interview that makes them look amateurish.
- Instrumental teacher who assigns an insensitive piece to their student without knowing it.

Lights, Camera, Action!

- Get to know someone in your field who you typically don't talk to. Ask them about their job, life, path or goals. Be genuinely curious.
- Cover the same song effectively for two totally different audiences.
- Clearly express the way you are feeling about a conflict, frustration, or disappointment in a healthy and direct way.
- Ask a friend how they are feeling when their face looks upset.

Social Intelligence VIP

Ara Guzelimian • Music Leadership

Ara Guzelimian has always held positions of leadership within the sphere of music. He has served as a dean, provost, artistic director, producer, host, consultant, and board member of some of the most prominent music schools and festivals in the world. In interviews, he speaks of the flexibility it takes to be in these various positions, where you oversee multi-faceted polices and are constantly navigating unexpected changes and challenges (Sosland, 2009). His social intelligence and experiences guide him through these decisions as he leans on the perspective of all parties at each level. He has an open mind about the fluid nature of what it takes to be a success in music. He also listens and takes time to understand the viewpoints of everyone that is involved. His illustrious career owes so much to his strong social intelligence and interpersonal skills.

References

HealthHub (n.d.). *6 ways to improve social skills and increase social intelligence.* Retrieved January 29, 2023, from https://www.healthhub.sg/live-healthy/584/mental_health_social_intelligence

Rezaei, A., & Bahadori Khosroshahi, J. (2018). Optimism, social intelligence and positive affect as predictors of university students' life satisfaction. *European Journal of Mental Health, 13*(2), 150–162. 10.5708/ejmh.13.2018.2.3

Rezaei, A., & Mousanezhad Jeddi, E. (2018). Relationship between wisdom, perceived control of internal states, perceived stress, social intelligence, information processing styles and life satisfaction among college students. *Current Psychology, 39*(3), 927–933. 10.1007/s12144-018-9804-z

Sosland, B. (2009, February). *A life devoted to the Arts.* The Juilliard School. Retrieved January 29, 2023, from http://journal.juilliard.edu/journal/life-devoted-arts

Sowell, E. R., Thompson, P. M., Holmes, C. J., Jernigan, T. L., & Toga, A. W. (1999). In vivo evidence for post-adolescent brain maturation in frontal and striatal regions. *Nature Neuroscience, 2*(10), 859–861. 10.1038/13154

Intelligent Change (n.d.). *What is Social Intelligence and why should I develop it?* Retrieved January 29, 2023, from https://www.intelligentchange.com/blogs/read/social-intelligence

JUSTICE STRENGTHS

Fairness

Musical Prelude to Fairness

Fairness is an aspirational goal for all music professions. There are entire industries built upon the principle of fairness – like music therapy which seeks to bring function, equity, and greater thriving to patients who have suffered a compromising injury, loss, or condition. Music education is another field that aims to bring knowledge to all children and, in more recent years, bring equity and diversity to its offerings so that music is represented on a more even playing field in the classroom. Fairness is a wonderful principle in the workplace where hard work and results yield advancement and reward regardless of race, gender, religion, orientation, or any other factor. There has been a push among the major music companies and organizations to have greater representation of diversity and gender at top-level positions.

All this structured and intentional fairness bumps up against a contradiction in the nature of life and the arts. Life isn't fair and, in the arts, one might witness the most disparity with outcomes not scaled based upon efforts. Not to say that effort isn't a huge part of getting to and staying on top, but it doesn't guarantee results as an artist. If you examine the dogged work ethic of the most famous musicians you know, you will see that they are always working, networking, improving, learning, and growing. Once you get the opportunity, the real work begins in order for you to hold onto it. There are also tons of other elements at play in the music industry including luck, chance, and being in the right place at the right time influencing success. You can increase your odds by working on your social intelligence, showing up and investing in your networks, and becoming knowledgeable about the forthcoming trends in both music and the business. But still, there are no guarantees in the music industry whatsoever – it's not a "one in, one out" contraption. In other words, the effort you put in is not congruent with an expectation of results. You can be the most amazingly talented singer who works every day on improving your technique and you might never get your big break. On the flip side, you might be a very beginning singer who picked up their phone and blows up on TikTok. This is not "fair," it just is.

There are conscious ways to include fairness and equity in the creative process of music-making. There are safeguards to protect your intellectual property and rights by way of split sheets, points on an album, credits, and all manners of contractual agreements ensuring fairness. In a group dynamic, it is important to establish these things early on, before something gets set in stone and you discover your co-creators are not on the same page about ideas and ownership. In other businesses, like operating a recording or teaching studio, it's important to share not only the profits but also the responsibilities. You must reinvest in those businesses by way of maintaining and repairing equipment and you also must treat those who work there with fairness and respect. And if you are lucky enough to have a manager as an artist, your part in fairness is to be accountable for your own behaviors. It's not your manager's fault if you showed up unprepared for a gig, session, photo shoot, or interview. Taking stock of your own behavior is a huge part of being fairer with others. If we are not self-reflective, we are developing fatal blind spots in our own practices.

What Is Fairness?

Fairness is a combination of voluntary actions and beliefs that confirm "other people matter." Their opinions count even if they differ from yours. Fairness is a moral compass involving logical reasoning – analyzing what is objectively right and wrong – and emotional reasoning based on empathy. As humans, we crave fairness for ourselves and others and usually have strong negative reactions when we experience or witness unfair treatment.

What does it mean to be fair? Fairness helps us to have a balanced view of things as it requires understanding the perspective of others. In different situations, there are different forms of fairness: sameness, deservedness, and need (Dobrin, 2012). Sameness means everything and everyone is equal. No one has more than another. Deservedness is the notion that you get what you deserve, and nothing is deserved if it isn't earned. Need is a feeling of common humanity that interweaves responsibility and compassion. Those who have should give to those who do not.

People who value fairness are more likely to engage in positive, prosocial behaviors like leadership or altruism. They do their best to avoid negative impact on others and can readily see another perspective. They use their logic to evaluate issues and have strong moral compasses. Being high in the strength fairness ties into self-evaluation and regulation. Growing fairness leads to higher personal well-being and mutually supportive relationships with those around us (Butorova, 2021).

Life and especially the arts present many scenarios which are not fair. This can sometimes feel overwhelming and hopeless. It is important to remember the perspective that you are not alone – all of us experience suffering and the randomness of life.

To gain more agency and peace, it is helpful to shift your concentration to areas in life where you have more control. From this vantage point, you can find the lesson in your experiences and put your energy into situations and people that help you grow. Then let the rest go! Just as it is good to have compassion for others' suffering, it is crucial to have self-compassion and focus actively on befriending yourself.

Fairness and Your Brain

Fairness activates the brain regions of the ventral striatum and ventromedial prefrontal cortex to help you be fair, and generous, and treat others the same way you would want to be treated (*Brain reacts to fairness as it does to money and chocolate*, 2008). Some might assume fairness is simply a ruse; something we adopt only when we see an advantage in it for ourselves. However, several journals report that the brain finds self-serving behavior emotionally unpleasant and finds genuine fairness uplifting (Kluger, 2013). All these emotional firings occur in fast and automatic bundles of neurons and brain structures. The brain favors fairness from the empathetic, emotional side over selfishness from the rational, survivalist side (*Are humans hardwired for fairness?*, 2008).

Affirmation

"I treat people the way I want to be treated."
"I give everyone a chance and don't let my personal feelings get in the way."

Quick Tip

Apply the same rules to everyone.

Musical Examples of Fairness

- Programming a concert with awareness of inequities
- Breaking up a solo to feature many members of the group
- Re-auditioning members of the ensemble mid-way through the year to allow social mobility
- Paying all genders equally
- Expecting all members of the band and crew to show up at call time
- Providing a sync opportunity to all artists you manage
- Practicing contract law
- Fighting for the underdog
- Insisting a class listens to soft a spoken student or quiet singer
- Hearing complaints with a sympathetic ear
- Fighting for inclusion and equality
- Going against corporations that overcharge for concert tickets
- Using music to advocate politically or socially
- Giving an unknown artist a shot

Discussion Questions/Self-Reflections

When was a time you remember when someone was fair with you in a way that really mattered to you? What did they say/do? What are three ways you used fairness this week?

_____
_____
_____
_____
_____
_____

Treble and Bass

Treble • Overuse of Fairness
PARTISANSHIP: prejudice, playing favorites for some people or issues you believe in and support while overlooking obviously unfair policies or treatment of others.

- Music chairs who only advocate for their own area without regard for the whole of the school.
- Music video director who takes over without regard for the artist's vision.

Bass • Underuse of Fairness
DETACHMENT: doesn't pay attention to injustices, disinterest, not caring, uninvolved in the whole group.

- Backing singer who loses interest in songs they are not dominantly heard on.
- DJ who doesn't care who is in the room and plays only what they want.

Lights, Camera, Action!

- Create a concert series for a marginalized population.
- Call upon your quietest student for their opinion.
- Look for a newcomer or someone who is not typically included in the group and find a way to include them in the conversation.
- Reflect on people or animals who are isolated, not included, or thought of with disgust. Be friendly to them and extend yourself to treat them fairly.
- Resolve a conflict with a partner or dear friend in a way that leaves you both feeling recognized, heard, and satisfied.
- Actively invite different opinions when solving a problem and keep the conversation open by welcoming all input and ideas.
- Consider ways to be fairer with friends and family.

Fairness VIP

Jordan Bromley • Entertainment Attorney

Fairness requires the discipline to see both sides of a situation in a measured way. People who exhibit a lot of fairness believe in treating people equally. Lawyers by nature must be able to use logic to assess multiple sides and points of view. They need, at times, to dispassionately weigh options and present their client with honest versions of possible outcomes. Jordan Bromely, top music attorney, speaks quite passionately about artists owning their art (*Jordan Bromley*, n.d.). His fairness is in play every time he is negotiating with a label or publisher on behalf of his artist client. He also exercised his fairness advocating for financial relief for pandemic-stricken entertainment workers under the CARES Act (Cullins, 2022).

References

Association for Psychological Science – APS (2008, April 15). *Are humans hardwired for fairness?* Retrieved January 29, 2023, from https://www.psychologicalscience.org/news/releases/are-humans-hardwired-for-fairness.html

EurekAlert! (2008, April 21). *Brain reacts to fairness as it does to money and chocolate.* Retrieved January 29, 2023, from https://www.eurekalert.org/news-releases/733977

Butorova, H. (2021, December 21). *The importance of: Fairness.* Citywise. Retrieved January 29, 2023, from https://citywise.org/importance-fairness/

Cullins, A. (2022, February 18). *From Beyonce's advisors to the lawyer behind Taylor Swift's masters acquisition: The Top Music Attorneys of 2021.* The Hollywood Reporter. Retrieved January 29, 2023, from https://www.hollywoodreporter.com/lists/thr-2021-power-lawyers-music/jordan-bromley/

Dobrin, A. (2012, May 11). *It's not fair! but what is fairness?* Psychology Today. Retrieved January 29, 2023, from https://www.psychologytoday.com/us/blog/am-i-right/201205/its-not-fair-what-is-fairness

Digital Entertainment World (n.d.). *Jordan Bromley.* Retrieved January 29, 2023, from https://www.dewexpo.com/portfolio/jordan-bromley

Kluger, J. (2013, October 4). *This is your brain on fairness.* Time. Retrieved January 29, 2023, from https://science.time.com/2013/10/04/this-is-your-brain-on-fairness/

Leadership

Musical Prelude to Leadership

Leadership and teamwork have a mutually beneficial relationship, although it is not always expressed with such reciprocity. Ideally, a leader respects and values the input of the team. They inspire the people they lead to come together and pool their energies towards a shared goal. The team in turn informs the leader of their needs. They are on the pulse of events and issues and are less removed from immediacy. This relationship can be quite harmonious!

Ideally, the team has chosen or elected a leader in whom they believe. In this very real way, leadership is a bestowed honor that comes with a lot of responsibility to the people they serve. Unfortunately, this balance is often thrown off by leaders who don't listen enough to the team or do things in a self-serving way, thus betraying that social contract. Their vision might be totally separate from that of the mission and the team's true heart isn't into making things happen with any expedience or worse, at all. The energetic undercurrent of the leader-team relationship is often at the heart of the success or failure of an endeavor, artistic project, educational institution, or music business entity.

Leadership is a privilege. Not only are you a figurehead but also a mentor, and you are being watched from many angles. Also, you can't choose to only lead during the good times and then recede and hide during the bad. The wins are not just yours; the losses are as well. Leaders often must take the fall for downturns, even if it was out of their immediate control or the team itself was responsible for something that went wrong.

In music, we have many leadership designations. There are infra-structures in education systems that dole varying degrees of power and agency – dean, chair, program director, conductor, principal, superin-tendent, etc. Bands have lead singers or guitarists who generally direct the artistic concept. Choirs have section leaders who guide the other voice-part members through learning the music. Orchestras have ranked and principal chairs who serve in leadership roles and have a special rapport with the conductor. The industry has all manners of leadership roles from CEOs to artistic directors to presidents.

Where there are leaders, there are followers who lean on the wisdom, ex-perience, and guidance of their chief. Akin to a head chef, too many cooks in the kitchen without organization or order can really confuse things. Ultimately, someone needs to assertively make the decision, choose the best option, or throw all their chips on the table. It takes courage, calm, and executive ability to be the person calling the ultimate shots.

If you happen to be a leader in your organization, engage your empathy, kindness, fairness, and justice. You would want to be led by someone who had an equitable moral sense. Also, a little zest, charisma, and authentic interest in the other go a long way in leadership. When people feel like you are genuinely engaged in them and their ideas, they authentically want to come over to your side.

What Is Leadership?

Leadership comes in many sizes and styles. We talk about "inspiring leadership" which goes beyond tasks and skills and addresses the whole person. There is a buzz surrounding these so-called "transformational leaders" (Cherry, 2022). This leader is genuinely passionate about ideas or goals and helps people feel included in the process. They are not solely concerned with the group or company growth but also cued into each team member's individual growth goals. There is genuine recognition, praise, appreciation, and sometimes rewards for people's accomplishments. However, perhaps the most important reward is feeling seen.

Transformational leaders tap into a deeper part of people and aspects of their meaning and purpose in life. They challenge people to bring their best and encourage creative solutions. They often change the landscape in their industry by allowing people to think and perform outside of the box. In music, we can see transformational leadership every time a convention is broken or challenged, and a new genre or convention is born. These leaders serve as passionate role models who walk the walk and talk the talk. People gravitate toward the enthusiasm and vision they can generate around a project or endeavor.

Leadership is a complicated orchestration of strengths and skills and aligns closely with fairness, teamwork, kindness, social intelligence, and gratitude. A considerable amount of money and training is dedicated to creating effective, good leaders as they are key to any team or company's success. Leadership has extensive research, supporting criteria, skills, and values.

There are many different leadership styles from authoritarian, to participative to delegative (Cherry, 2022). It is important to consider which style or combination of styles you fall under. Taking the VIA character strengths survey is a great place to start! Know how you lead and think and identify your strong and weak areas for your ongoing personal and professional journey.

All leaders can encourage divergent creative thinking by accepting input and allowing time for their team to think of as many possibilities as they can before narrowing down choices. Don't kill an idea before it has a chance to be born. When someone shares, receive it with curiosity and appreciation – it is a gift they are offering.

It's beneficial to have a positive, energetic, and optimistic disposition when you are leading a team. People are inspired by those types of personas and will be more willing to follow them. Modern leadership is a co-created collaborative process in place of a top-down model (Rosenkranz, 2023). Leadership is best seen as a mutual learning process and the growth goes in both directions – between the leader and the people they lead. It can be very exciting if you are open to it.

Leadership and Your Brain

Having this character strength strongly correlates to lower levels of depression and anxiety. The basis of leadership skills is born in your brain. The right frontal area of the brain is a crucial center for interpersonal communication, social skills, and relationships (Nelson, 2022). Understanding how brains are motivated is helpful to leaders.

Employers who only use financial bonuses as motivation are missing out on all the intrinsic aspects of human nature.

Research shows that our brains are social organs, hardwired for connection. A study at Case Western Reserve University discovered a relationship between empathy/resonance and effective leadership (Boyatzis, 2015). The ability for a leader to put themselves in the shoes of those they are leading is key. Social intelligence goes a long way when stepping onto the conductor's pulpit, sitting at the head of a boardroom table, and looking into the crowd then back to your band and calling the next tune.

Resonant leadership styles also help build trust through a mutual release of oxytocin (Zel, 2019). In contrast, a dissonant style is more authoritarian and objective. Fourteen regions of the brain responded when thinking of resonant leaders, while only six responded when thinking of dissonant leaders (Nelson, 2022).

When you are designing or tweaking your own leadership style, keep in mind your intended result. If you want increased attention and positive relationships, consider being a more resonant leader.

Affirmation

"I like to organize activities for others and get things done with harmony."
"As a leader while I like to get things done, it is important to me that I make everyone feel included."

Quick Tip

Find your own style of leading – be natural, honest, and authentic.

Musical Examples of Leadership

- Organizing a huge musical conference and delegating tasks
- Recognizing secondary instrumental ability in other members of the band
- Managing a roster of artists
- Starting a music blog
- Operating and managing a recording studio
- Running a music lesson business
- Confidently calling out tunes in the set
- Taking control of an out-of-control music classroom or assembly
- Directing a team of interns
- Organizing union activity
- Establishing the creative direction of a video shoot
- Forming a group for women engineers
- Controlling your own branding decisions as a pop artist

Discussion Questions/Self-Reflections

What kind of leaders or leadership styles do you respond best to?

Describe the best leadership experience you have had and the things the leader did to make it so.

_____
_____
_____
_____
_____
_____

Treble and Bass

Treble • Overuse of Leadership

DESPOTISM: dictatorial, you know what's right and everything must be overseen by you.

- Marching band director who relentlessly rehearses their band in dangerous weather conditions.
- Producer who demands all the credit for a project they were barely involved in.

Bass • Underuse of Leadership

COMPLIANCY: just go with the flow, it doesn't matter to you, disengaged, someone who just shows up.

- Lead singer who lets the guitarist dictate artistic direction although it goes against their values.
- Music teacher who changes a student's grade for fear of administrative pressure.

Lights, Camera, Action!

- Host a monthly social gathering where you can exchange ideas with colleagues.
- Raise your hand to run a sectional or rehearsal.
- Organize and lead a group around a cause you want to support.
- Have a standing office hour open to employees specifically to share their ideas about how to improve the work environment.
- Discuss ways an employee can incorporate their signature strengths into their work life.

Leadership VIP

Valerie Coleman • Composer

Leaders are not often the ones announcing themselves as the head of a table, as the cliche might have one think. Real leaders know the formula for making something run smoother, keeping the entities involved in sync, and recognizing excellence in the others around them so that their ultimate vision is realized. Dr. Valerie Coleman is such a leader on many levels (*Valerie Coleman flutist & composer*, n.d.). As a Grammy-nominated artist, and university professor specializing in flute, composition, and chamber music, she literally leads students and directs them towards not only artistic success but encourages entrepreneurial success. Her 24-year engagement with Imani Winds as a flutist and founder exemplifies her leadership in both business and mentorship. And as an in-demand modern composer, she has been the first living African-American female composer commissioned by the Philadelphia Orchestra at Carnegie Hall. She also leads as a black female composer in a world dominated by white men with authenticity, grace, and power. Leaders such as Dr. Coleman inspire the future generations of composers to live, compose, play, and speak their own stories and truths.

References

Boyatzis, R. (2015, February 13). *Neuroscience and leadership: The promise of insights.* Ivey Business Journal. Retrieved January 29, 2023, from https://iveybusinessjournal.com/publication/neuroscience-and-leadership-the-promise-of-insights/#.UwZW7fldVSB

Cherry, K. (2022, May 23). *How to become a stronger and more effective leader.* Verywell Mind. Retrieved January 29, 2023, from https://www.verywellmind.com/ways-to-become-a-better-leader-2795324

Cherry, K. (2022, November 14). *6 Leadership styles and Frameworks.* Verywell Mind. Retrieved January 29, 2023, from https://www.verywellmind.com/leadership-styles-2795312

Nelson, A. (2022, November 29). *The basis of leadership is born in the brain: Neuroscience in business.* Hppy. Retrieved January 29, 2023, from https://gethppy.com/leadership/basis-of-leadership-neuroscience

Rosenkranz, J. (2023, January 10). *The Future of Leadership: Co-creation is replacing top-down management.* Howspace. Retrieved January 29, 2023, from https://howspace.com/blog/the-future-of-leadership-co-creation-is-replacing-top-down-management/

Valerie Coleman Flutist & Composer (n.d.). *Valerie Coleman flutist & composer.* Retrieved January 29, 2023, from https://www.vcolemanmusic.com/

Zel, U. (2019, August 21). *Neuroscience of Leadership.* LinkedIn. Retrieved January 29, 2023, from https://www.linkedin.com/pulse/neuroscience-leadership-ugur-zel/

Teamwork

Musical Prelude to Teamwork

Every music profession requires great teamwork. Even the most solitary of solo artists can only achieve greatness and success by engaging a team. There are many points along the journey that fall outside of the artist's expertise, experiences, connections, purview, or abilities. Think of all the separate logistics of booking a tour, collecting money, filing taxes properly, renting the correct sound and lighting equipment, scouring the contracts, promoting the project, and producing, mixing, and mastering the album. The 'do it yourself' model is only true to a degree. At some point along a successful journey, you will want to hire people who are experts in these separate areas. You can't really be the jack of all trades and do everything yourself – at least not well. Why would you want to anyway? You may dabble in promotion an hour of your day, but there is someone who does this eight of theirs. Learning how to cultivate and work with experts in their own field is an amazing and important part of growing in your career.

This doesn't extend only to performers. All aspects of the business side of music work with teams of people who niche out their skill sets to leverage the best results. Communication is an essential element of great teamwork. This involves honesty and knowing your strengths and weaknesses. The character strengths assessment is a wonderful tool to start with when joining or forming a new team. Once you begin to know each other in this way, you can delegate tasks in a more natural manner. When you can engage people in their strengths, there is a joyfulness to the work that makes it lighter and more in flow with the person's abilities.

The creative process also can heavily involve teamwork. At the top levels of songwriting, you may be one of many in a room working towards the one goal – the best hit song you can write. There is no captain in this scenario and individual creators must learn how to work together in a team. It's a true group effort and sometimes that can be tough to master. Many creators feel like they would rather be alone on an island than compromise their vision to a roomful of other creators. However, it is very important to learn how to play with others as a writer, performer, educator, industry person, and any other trade or profession where you are not paid to sit solo on your island.

It's rewarding to learn how to let go of an overly controlling tendency. You might like not having to be the one-person-show handling everything. Knowing and recognizing when you need help is essential in this. Asking for help is the next step. You can optimize your efforts by learning where your talents and time are best placed. You may even find that once you are en-sconced in a team of people you really enjoy who make your life easier, you never again pick up the paddle to go back to island living.

What Is Teamwork?

Witnessing good teamwork is like watching a winning athletic team play, a graceful dance troupe perform, or a beautifully directed symphony. There is a balance of synchrony and diversity. Everyone is committed for the good of the team and people put their individual needs and egos aside to reach the team's goals. In these scenarios, it is easy to see the high-quality re-lationships between individuals and the respectful way they speak to and treat each other. You see and experience this through the process and the production.

The highly functioning team has a palpable spirit, drive, and energy. The group culture is supportive and affirming and group members respect each other and their contributions. There is an open-minded atmosphere that is inclusive and honoring. People feel expanded and creative as part of this team and learn as much as they contribute. This type of team is a win-win growth experience. Also, this kind of support creates openness (Andreatta, 2020). Harvard professor Amy Edmondson defines this as workplaces that have psychological safety (TEDxTalks, *Building a psychologically safe workplace*, 2014). One must feel they can speak up with their ideas without fear of being shut down, retaliated against, or belittled. Organizations can quickly become toxic when leadership is an agent of suppression and decimation of the individual voice. It is very important to listen to the concerns of your employees at every level – they are much more in touch with the day-to-day challenges than someone who sits at the top leading only with ideologies.

Some important team perks are deepening levels of social trust and positive views of human nature. Teamwork also improves productivity as individuals are more engaged and motivated to reach team goals. Great communication between team members minimizes friction, errors, and delays (*Top 10 benefits of teamwork you must know!*, 2022). You also can learn from the mistakes of others and work together to find a solution. This brings people closer together and opens more vulnerability in positive, socially bonding ways. Think of a time when a lack of team communication caused issues in your work environment.

A diverse team working together creates a greater synergy – the whole is greater than the sum of its parts, that is the totality of individual efforts. There may be no better example of this than musical groups – whether it is an orchestra – full of all types of instruments and sounds coming together under leadership and sheet music to make something striking, or a rock band aligning to the drumbeat, the lead singer, and the energy of the crowd to make something show-stopping. It is easy to see how music is almost always a team sport. Even a solo artist relies on team members to round all the efforts of their career – from recording their album, booking their shows, marketing them, and collecting royalties. The clearer and smoother the communication and teamwork, the better the results.

Diversity is a strength when forming a team. Even though it is sometimes easy to believe you would want a bunch of "yous" working on your team, would you really? Your own blind spots would be magnified, and your own strengths would be futilely duplicated. Having a multitude of differing perspectives, ages, cultural backgrounds, and experience levels ensures a wider view of any one problem or task (Haas & Mortensen, 2022). Think about how diversifying strengths on your personal team would help you achieve your goals. Or think about how your team member's complementary strengths can really fill in gaps to make the whole. The front-facing talent needs a hearty back office or there will be elements not handled. These two types may have completely different strengths. Learning to hire and partner with people who can fill in your blank spots can create amazing results! Effective teamwork requires a mindset of commitment; "I am in it for the long haul and the people I am working with are as important to me as completing the task at hand."

While the pandemic realities prevailed, our technology made it possible to work with others through computer screens and phones. This was crucial in keeping us connected during difficult and lonely times. Let's remember though, our brains were built for in-person interactions as they decipher meaning and intent through tiny micro-muscular facial changes, body language cues, and pheromone signals (Andreatta, 2020).

As we transition forward to a new normal, we need to remember and honor the power of in-person interactions -where the maximum number of cues are available. And if you have participated in pandemic-era "Zoom happy hours," virtual birthday parties, and drive-by events, you know this doesn't substitute for the joy of getting together in person with your team.

Teamwork and Your Brain

Our brains are hardwired for connection, helping us create meaningful bonds and being able to discern positive from negative or toxic connections (Levbre, 2017). Discoveries in neuroscience can guide us in building and managing more positive teams and groups. The challenge is to activate brain structures for trust and collaboration versus conflict and competition. Our brain's response to the latter dulls and decreases our ability to perform at peak. It was also discovered that being excluded from a group lights up the same brain regions as actual physical pain (Andreatta, 2021).

Team flow is another example of a complex orchestration of brain activity. When one is "in the zone" of being on a team, the brain creates a unique signature of flow with more synchronized activity among teammates and increased beta and gamma brain waves in the middle temporal cortex (Shehata et al., 2021).

It is even suggested that teamwork was a core factor in humans developing big brains! This draws from the social intelligence hypothesis which explains that we need these large brains to keep track of the complicated cast of characters in our every-moment interactions – who is friend, who is foe, who is a social rank above us, etc. (Fields, 2012). Parallel to this idea, some studies show that bigger-brained primates tend to live in bigger social groups (MacLean et al., 2013). The strength of teamwork is associated with "sustainable behavior." This is behavior that protects the social and physical environment for survival, consistency, and in our context, thriving and finding the good in each other and life.

Affirmation

"I love to work side-by-side with others."
"I am a loyal and dedicated team member."
"I do my share and work hard for the success of the group – it's not about me!"

Quick Tip

We're all in this together!

Musical Examples of Teamwork

- Volunteering for a musical charity
- Working without ego with your team members towards an artist's release
- Delegating teaching responsibility to section leaders in a choir
- Rolling up your sleeves and helping with all tasks in a large event
- Conscientiously getting the lunch order right as an intern at a famous recording studio
- Striving for a perfectly blended sound in a choir
- Finding a rhythm working with the crew for an arena concert
- Equally promoting band members' musical projects
- Partnering with other businesses to put on a webinar
- Learning to ask for what you need
- Helping to break down equipment and wrap cables after the show
- Setting up chairs for the orchestra before rehearsal
- Recruiting for your school or at a college fair

Discussion Questions/Self-Reflections

Here are the main ingredients for a successful team: Communication, commitment, goals with plans, support, delegation, accountability, and respect. Think of a team you are involved in currently. In which ingredients are you most strong? Which do you need to grow? Pick one of your growth strengths and write out two ways you can work on it with a team you are involved with.

_____

_____

_____

_____

_____

_____

Treble and Bass

Treble • Overuse of Teamwork

DEPENDENCY: mindless and automatic obedience, cannot do things without everyone's approval or okay, stifling own creativity.

- Artist who overly relies on the opinion of their social followers before they can make an artistic decision.
- Performer who doesn't read music and relies on someone to transcribe their own songs.

Bass • Underuse of Teamwork

SELFISHNESS: rebelliousness, self-absorbed. It's easier to go it alone and not deal with other people's issues. It's OK to sit back and allow others to do the work.

- Accompanist who makes their playing all about them and not who they are accompanying.
- Guitarist who takes every solo without regard to the rest of the band or the song.

Lights, Camera, Action!

- Take on a larger project with several others: composing music for a play, submitting chapters for a book, marketing an album.
- Start a band!
- Lean into a challenge that your organization is having and take some ownership of being part of the solution.
- Express appreciation for someone's contribution.
- Ask a teammate for feedback about you on a recent project
- If you are in a relationship, describe how you and your partner act as a team together to deal with the world and solve problems.

Teamwork VIP
Rick Rubin • Producer • Engineer

A person strong in teamwork believes in the "whole" – whatever that whole may be: a sports team, a relationship, a band. Influential and seminal producer, Rick Rubin, is one of those ethereal musical team members who is "all in." His ability to enter the aura of a project, push the artists to explore and go deeper, leave room for the magic and egolessly guide musicians along their journey shows exemplary teamwork (MacLean et al., 2013). His peaceful, receptive, and present demeanor creates an atmosphere that allows space for the artist to come out of their own subconscious longings (*Pharrell and Rick Rubin have an epic conversation* 2019).

References

Andreatta, B. (2020, November 18). *Better together: The neuroscience of teams: ATD.* Main. Retrieved January 29, 2023, from https://www.td.org/insights/better-together-the-neuroscience-of-teams

Andreatta, B. (2021, April 29). *Peak-performing teams need psychological safety.* Britt Andreatta. Retrieved January 29, 2023, from https://www.brittandreatta.com/exclusion/

Blog.bit.ai (2022, June 20). *Top 10 benefits of teamwork you must know!* Bit Blog. Retrieved January 29, 2023, from https://blog.bit.ai/benefits-of-teamwork/

Fields, H. (2012, April 10). *Teamwork builds big brains.* Science. Retrieved January 29, 2023, from https://www.science.org/content/article/teamwork-builds-big-brains

Haas, M., & Mortensen, M. (2022, March 30). *The secrets of great teamwork.* Harvard Business Review. Retrieved January 29, 2023, from https://hbr.org/2016/06/the-secrets-of-great-teamwork

Levbre. (2017, April 9). *Hardwired for connection: Therapy & marriage counseling.* PsychCare. Retrieved January 29, 2023, from https://www.psychcaremd.com/hardwired-for-connection/

MacLean, E. L., Sandel, A. A., Bray, J., Oldenkamp, R. E., Reddy, R. B., & Hare, B. A. (2013). Group size predicts social but not nonsocial cognition in lemurs. *PLoS One*, 8(6). 10.1371/journal.pone.0066359

Mobbs, D., Hagan, C. C., Dalgleish, T., Silston, B., & PrÃ©vost, C. (2015). The ecology of human fear: survival optimization and the nervous system. *Frontiers in Neuroscience*, 9. 10.3389/fnins.2015.00055

Shehata, M., Cheng, M., Leung, A., Tsuchiya, N., Wu, D.-A., Tseng, C.-huei, Nakauchi, S., & Shimojo, S. (2021). Team flow is a unique brain state associated with en-hanced information integration and Interbrain Synchrony. *Eneuro*, 8(5). 10.1523/eneuro.0133-21.2021

YouTube (2014). *Building a psychologically safe workplace*. YouTube. Retrieved January 29, 2023, from https://www.youtube.com/watch?v=LhoLuui9gX8

YouTube (2019). *Pharrell and Rick Rubin have an epic conversation*. YouTube. Retrieved January 29, 2023, from https://www.youtube.com/watch?v=PnahkJevp64

First Set Practice Room

Tell a best-self story. This is a story about a time when you felt you were at your best or were proud of something you did or felt particularly in "flow." Speak this out loud, record it, or write it down. You will then strength-spot by making a list of all the strengths you heard. The character strengths cards are a great tangible and visual tool to use for this activity (Merch Booth). You can also do this activity with a partner and have them list all the strengths that showed up in your story.

CHAPTER 3

Second Set

Explore – Virtue Categories and Music Professions

Now that the reader has a deeper understanding and context for their own signature strengths, we will explore the larger virtue categories to which these 24 strengths are organized.

Six core virtues emerged from Peterson and Seligman's research on positive psychology that were present across cultures and time: **courage, justice, humanity, temperance, transcendence,** and **wisdom.** The assignment of the strengths to the virtue categories, as seen in Figure 3.1, was based on theoretical grounds.

The virtues are groupings or clusters of different kinds of strengths that describe the true nature of that strength and include courage (emotional strengths), temperance (protective strengths), wisdom (cognitive strengths), transcendence (spiritual strengths), humanity (social strengths), and justice (community strengths).

What followed next was Peterson and Seligman's proposed model of classification that includes both horizontal and vertical dimensions. It is sometimes said that what resulted was like a handbook of mental wellness and a good thriving life ("Values In Action Inventory of Strengths," n.d.).

You will see how the demands and job descriptions of the different music professions fall under each general category. Note, these are suggestions for each profession and a case could be made for a profession to fall under several categories as strengths work together. The spirit of each example seems to gravitate to a certain virtue category.

Lastly, you will read case studies of music professionals and how they leverage their individual strengths for success in their field. Though their profession was chosen for the virtue category, you will see how their unique strength profile infuses the way they perform their jobs. You will see the various virtue categories that make up their signature strengths.

DOI: 10.4324/9781003267607-3

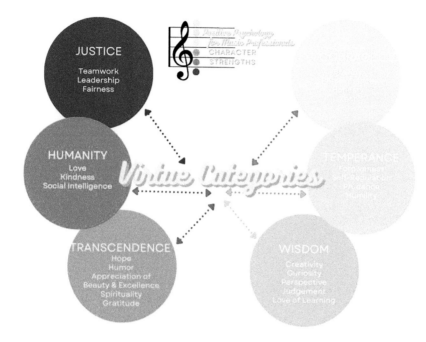

FIGURE 3.1 Virtue Categories with All 24 Strengths

At the conclusion of this chapter, you will be guided to practice room three to explore aspirational people and examine the strengths they exhibit.

STRENGTHS TRIAD

Triad

1. *noun*: Triads are made up of three notes played together. You will often hear people refer to triads as chords. They consist of a bottom note (the root), a middle note (the third), and a top note (the fifth).

In music, a three-note triad, or chord, is made up of a low, a middle, and a high note. Musicians often refer to the low note as the root, the middle note as the third, and the highest note as the fifth. The third and fifth are labeled for their intervallic distance from the root. In Figure 3.2, you will see a G major chord.

We are going to use a musical analogy here to give a visual parallel. In the same way these three notes create harmony, you can think of your strengths

FIGURE 3.2 Musical Triad (G Major Chord) Showing the Root, the Third and the Fifth

being arranged to create your "harmony" or your distinct personality. Our 24 strengths can be grouped into three groups of eight:

- Your highest ranking eight (the ringing highest note of the chord)
- Your middle ranking eight (the middle note of the chord)
- Your lowest ranking eight (the lowest note of the chord)

The rank order of these creates your unique strengths profile. See the strengths in Figure 3.3 grouped in a triad formation with the eight highest (the fifth of the chord), eight middle (the third of the chord), and eight lowest (the root of the chord).

FIGURE 3.3 Strengths Triad Grouping High, Middle, and Low Strengths

It is useful to take a wider view of your top strengths to get a larger picture of yourself, especially when looking at the virtue categories. In looking at your top eight, do they favor one category over another? Are they split between two categories? Are they equally spread out?

Let's look at a sample of strengths in this triad formation (Table 3.1):

Table 3.1 Strengths Triad Example

Ranking	Strengths
HIGHEST 8	Creativity, Appreciation of Beauty & Excellence, Zest, Humor, Curiosity, Love, Gratitude, Perspective
MIDDLE 8	Spirituality, Love of Learning, Fairness, Social Intelligence, Honesty, Hope, Bravery, Kindness
LOWEST 8	Perseverance, Leadership, Forgiveness, Self-Regulation, Humility, Prudence, Judgment, Teamwork

Now take a moment and fill this out for your own strengths. You can refer to the STAGE PLOT section where you may already have written your HIGHEST 8 and to the results of your VIA survey for the rest (Table 3.2).

Table 3.2 Your Strengths Triad

Ranking	Strengths
HIGHEST 8	
MIDDLE 8	
LOWEST 8	

In this chapter, we ask you to look at your HIGHEST 8. You may find it useful to look at this broader view of your strengths in the context of looking at the categorized music professions. How do your high eight strengths show up in your current profession, activities, or studies? Look for everyday simple examples and/or larger patterns.

_____

_____

_____

_____

_____

_____

_____

_____

As you read this chapter, pay attention to the professions that seem to require the strengths you possess. You may discover new interests or un-thought-of career paths by following your strengths!

STRENGTHS IN EACH VIRTUE CATEGORY
Courage

FIGURE 3.4 The Courage Strengths: Bravery, Honesty, Perseverance, Zest

The Courage Category

Courage is all about powering through to reach your goal. Perhaps you are finding your place in the music business, attaining education, breaking a glass ceiling, or taking artistic risks. On any of these journeys, you may encounter physical, emotional, and/or mental resistance. The strengths in this category are all about tenacity and endurance through barriers despite your fears and resistance (Figure 3.4).

This is expressed differently ranging from personalities that boldly go after things head-on to those who persist with introverted, quiet courage. People tend to glorify the first example but often our greatest heroes are unseen, quiet, and strong individuals. We can be courageous in both *fortissimo* and *pianissimo* dynamics. Whichever manner is externalized in a person, they both first happen internally.

And don't confuse having courage with fearlessness. Courage is the tool that allows you to proceed despite your fear. Perhaps the most courageous of us are concurrently the most fearful. Experience is the greatest teacher, and we all develop grit to get through life's hardships. It is important to log life successes, by journaling and talking about them, to help you put your fear in a healthy perspective.

Courage is an important category of strengths when it comes to expressing individuality and speaking one's truth. It may seem "easy" to be oneself but in this day and age of social media, there is an unspoken pressure to conform. This happens personally and musically whether for acceptance, taking the path most traveled, for views and clicks, or to 'fit in."

This category of strengths also can serve as a great "battery" to energize all your endeavors, injecting joy and spirit into your daily grind. There is a vibrant and buzzing energy to the courage category that can be quite inspiring and contagious!

Temperance

FIGURE 3.5 The Temperance Strengths: Humility, Forgiveness, Self-Regulation, Prudence

The Temperance Category

Temperance is all about keeping yourself in check. Though the strengths in this category don't appear flashy and loud, they are perhaps some of the most important superpowers you can have to maintain positive relationships with others and yourself. These strengths form the backbone for healthy discipline that all careers require (Figure 3.5). Lastly, they also invite a beautiful pause for self-reflection.

The music industry has a front-facing, achievement, and metrics-driven side which tends to get all the airtime. However, analyze the career of anyone with success in this industry and you will find a ton of self-regulation, discipline, and prudent business choices. These slower-tempo strengths create a scaffolding to help you carefully plan and execute your goals with purpose. The act of achieving virtuosity in music – as a violinist, producer, researcher, or industry mogul – requires hours, days, and years of self-regulation, reassessment, and constant growth. The greats on top of their game have this mentality, allowing them to transform with the ever-changing landscape of music. It's what keeps them on top. They are never "done" evolving, growing, and grinding, regardless of the high places they have reached.

Forgiveness and humility are grounding, of-the-earth strengths – they stabilize us. Humility gets a strange reputation when it comes to the arts. It's more easily seen in selfless or service roles and less in the hotter spotlight. However, when operating from a place of humility, you allow yourself to check your own ego – or the assumption that your way is the best way. When you can loosen yourself from this, a world of other options can open. Also, a great deal of work in this profession requires close-knit teamwork. Humility allows you to listen to others' opinions equal to your own and stay open. Humility allows you to see the genius and creativity in others, providing you with one of the greatest gifts – to becoming a lifetime learner.

Forgiveness is an important ingredient for healthy relationships with others. Forgiveness is not an act of weakness or endorsement of bad behavior. It's the recognition of our common humanity and faults and a decision to move forward intact while staying connected with others. Holding grudges, like holding anything heavy, requires your energy, effort, and time. When we focus negative energy intensely on an unpleasant experience or person it is equal to giving away your energy, passion, and focus. People who refuse to forgive are actually manifesting more internal connections with that bad scene. Wouldn't you rather be proactively manifesting and cultivating an experience you *want*? Forgiving releases people from your energy field and allows everyone to move on.

And forgiveness is not just a gift you can give to others, it is perhaps most powerful when directed at yourself. Self-compassion is an expansive growth strength. The energy of holding a grudge against yourself is energy that is better aimed toward growth. Take the lesson and learn it, don't punish yourself with it. Look at yourself with soft eyes and a warm heart and realize we are all humans with limitations.

When we choose forgiveness, we get that energy back to do what we wish!

Wisdom

FIGURE 3.6 The Wisdom Strengths: Love of Learning, Perspective, Curiosity, Judgment, Creativity

The Wisdom Category

Wisdom strengths have a divine intelligence to them. They are a deep gaze into the crystal ball of "why," "how," and "what if." They take us on a journey with an unknown destination. These strengths drive forward change, invention, new genres, and technology while glancing at the rearview mirror of knowledge and perspective (Figure 3.6).

The vanguards of all music professions move and shake from the impetus of wisdom strengths. Curiosity is a dare to the convention. Creativity is bold action against fear. The love of learning fertilizes new growth, ideas, and domains. And judgment and perspective inform the whole while drawing upon the whole breadth of experience.

Strengths in this category are all about invention and reinvention. Artists use creativity, curiosity, and judgment on their path to finding their style or sound. The music tech industry is fueled by futuristic creativity driven by curiosity and informed by the perspective of the past. Entertainment attorneys literally use perspective, judgment, and knowledge as they parse contracts or prepare copyright infringement cases.

There are many times in life when artists feel creatively "stuck." Strengths in this category can help them explore undiscovered musical terrains with openness and excitement, rather than self-judgment and fear.

Artists are not the only ones who come to these crossroads. The "ever-under-construction" nature of the evolving music industry has forced its members to sometimes change entire fields as new budgetary models, technology, and trends are built and rebuilt. The scariness of these unknowns can be lessened and better navigated with a good dose of curiosity and love of learning. These strengths can help you investigate other areas of the industry you don't even know you have an aptitude or passion for. Some of these areas are emerging as this is being written! Let the wisdom strengths guide you to discover parts of yourself you have not even met yet.

Transcendence

FIGURE 3.7 The Transcendence Strengths: Hope, Appreciation of Beauty and Excellence, Gratitude, Humor, Spirituality

The Transcendence Category

Music itself is transcendent. It feeds the soul. Some of the most difficult moments of our lives are helped by music. It is described by some as a universal language. Woven into it are magical golden emotional threads that have the power to both move us to tears and exhilarate us to heights unknown. Music is a beacon of hope, a celebration of beauty, a spiritual experience, and totally personal to the receiver.

The strengths in this category often naturally resonate with those involved in the musical fields as they reflect the spirit of music itself (Figure 3.7). They may represent the more sensitive parts of you or the siren that first called you to serve music. Take a moment now and recall – was there an influential moment of awe, hope, spirituality, or recognition of beauty on your personal path that led you to where you are now?

This category is all about connection to something ephemeral, meaningful, and unifying. It's getting outside of your own body and linking to something higher – a religion, a feeling, a common experience, nature, a belief, or even a piece of music itself. Think of the power of a national anthem played at an important sporting event or at the winners' podium of the Olympics.

Humor and gratitude are tools that serve to keep you balanced and light on this earth as life throws its punches. They keep you both agile and rooted. Laughter is medicine. It allows sunlight to beam through the darkness. Gratitude is the queen of all virtues as it offers a perspective that keeps you feeling blessed by finding the good in all situations. When things go awry, keep these strengths close to the heart.

Hope and spirituality can go hand in hand. To believe in something greater than yourself is to open to the hope of possibility. Depending on the specific job, professions in music tend to lean toward the winding path. There certainly is no "sure-fire" formula towards stardom – and if there was, that would be the subject of the next book! The music industry undergoes huge waves of change based on technology and the ever-evolving ways people relate to and consume music. These rough seas swallow up entire parts of the industry, making them obsolete while also birthing completely new areas that require new skills and mastery. Hope is essential on your bumpy sail toward the horizon as you learn to adapt and navigate these changes.

Spirituality connects with music quite literally. Think of all religions and spiritual practices. Music and sound are inextricably tied to so many of their rituals. Why is that? Music tends to unlock our emotions. Put your favorite movie on mute during its climactic scene and see how so much of the emotion is wicked away. Music also gives us a common experience. Whether a hymn, a chant, or a singing bowl, it brings us into the same frequency. Spirituality is also a grounding, perspective-bringing strength that helps us endure challenges and celebrate the good.

And last while certainly most, appreciation of beauty and excellence. It's the ability to momentarily float in a state of total awe. This sacred strength lets you fully savor the fruits of life. Celebrating the richness around you allows you to put on those beautiful rose-colored glasses and take in the good and fill up your positivity tanks to call upon and savor in the future. And this magnifies the magic you feel in your own affairs.

Appreciation of beauty and excellence extends also quite literally to creating, coaching, or acknowledging virtuosity. The loving attention one must place on the most mundane parts of learning music is nourished by this strength. A classical guitarist might practice their piece well over 100 times, sometimes taking each phrase in slow motion so that the brain and the hands create the exact math of each harmonic and melodic shift. This requires stamina, patience, and hope in service of that goal of beauty and excellence.

Mixing and mastering engineers go through a similar process in micro-analyzing elements of music that perhaps most people can't even hear! But those delicate, delicious details make the magic.

And at the heart of this push towards excellence, most people can fondly remember a teacher who believed in their excellence and potential. All teachers have the big-picture assignment to see the statues within the blocks of stone. They are called to cultivate the flowers of excellence in their students in a cyclical garden that grows and changes every season. Perhaps there is a teacher of yours you would like to express your gratitude towards through a letter or an actual visit (Tomasulo, 2011).

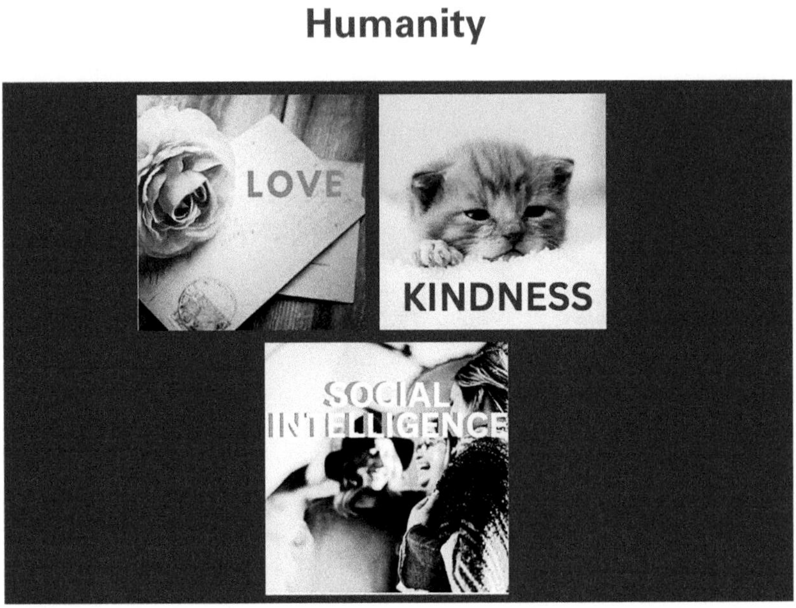

Humanity

FIGURE 3.8 The Humanity Strengths: Love, Kindness, Social Intelligence

The Humanity Category

These are your warm and fuzzy strengths (Figure 3.8). They involve sensing the needs and energy of others. Humanity strengths activate your empathy and generosity. They allow you to give from a pure place of spirit, without the expectation of reciprocation. At the core, they seek our commonalities. Empathy is the ability to sense other people's emotions, along with the ability to imagine what someone else might think or feel. Empathy allows us to recognize that we are all part of the human or animal condition. When we can see ourselves as part of this greater whole, we can act with more compassion and self-compassion, even when we feel hurt and wronged.

Love is such a powerful emotion and there are different kinds of love. We are bound to each other by love, and we pursue our passions with our love. We have love for our family, friends, partners, pets, and professions. Falling in love with someone is one of the most chemically intense experiences you can have, and heartbreak is one of the most painful. The subject matter of so many songs centers around these two love experiences. Love is also what keeps us on our music grind when there may be so many other options that would be safer and easier bets. Love reminds us of who we are at our core and how we connect to others. Love is a super strength that can get us through difficulties and can give us hope for better days ahead.

Kindness is the gentle spirit of love in action. To be kind, one must be mindful and consider the feelings and experiences of others. Kindness is a beautiful act done for the sake of itself. It doesn't seek credit. It's altruistic and intrinsically rewarding. There are musical charities that many of you may be involved in whereby you give your talents and resources for the sake of kindness. It also connects to the transcendent strength of gratitude.

Social intelligence comes in many forms. It can look like the producer who knows exactly what the artist in front of them is wanting to say but can't. It can be the A&R (artist and repertoire) person who understands why a crowd is connecting to a band. It can look like the teacher who pays extra attention to the introverted student. It can be the job candidate who highlights the perfect example of their experience and finds personal avenues to relate to their interviewer. Or it can be the musicologist who is able to investigate archives and detect different themes relating to their subject's life or humanity.

Whichever way it's expressed, social intelligence helps everyone in all situations. To put it in another way, think of a time when you worked with or for someone without this strength. Perhaps they tended towards narcissism and were incapable of seeing past their own needs or experiences. Was that a difficult relationship? Social intelligence can be a superpower when working in teams, bands, or classrooms (Figure 3.9).

Justice

FIGURE 3.9 The Justice Strengths: Fairness, Leadership, Teamwork

The Justice Category

Justice strengths are about the greater community (Figure 3.9). They have equity at their core. Oftentimes we are struck by the many ways life is not fair. These strengths aim to right that by allowing all voices to be heard. Fairness is a subject that comes up a lot in the music industry – whether we are discussing copyrights, royalties, contracts, streaming, song splits, etc. It comes up in matters of programming concerts to represent minority and women composers. We also deal with fairness when auditioning parts in a school musical. It comes into many of our decisions in music. We have invested resources into bringing greater awareness in terms of inclusion and equity to help bring more fairness into institutions and performance arenas. Those who exhibit fairness in their signature strengths may be seen as warriors by their peers as they can't stand to see injustice done. They will put in the extra work to help bring things to balance and will do this from a fiercely strong inner place.

Leadership is a justice strength when it is done with little ego. We probably all know leaders who are selfishly motivated or those who seek titles and power for the sake of increasing their self-worth. These leaders often miss the honor and privileged part of leading that serves the greater good. Great leaders are always thinking about the health of their organization, and they pay attention to the humans who will face consequences for their decisions and actions. They are thinking both horizontally and vertically, looking at the past, present, and future. These evolved leaders earn our trust. This can happen at any level in music – in a school, in a rock band, or at a record label. When a leader truly is on the pulse of their constituency, they can achieve great things because they will have the power of the people authentically behind them. They don't rule with fear tactics but rather with altruism – they work to bring out the best in their people.

Teamwork is fundamental in so many ways within music. A string quartet is a four-person team who must constantly listen to each other and respond musically. A record label is a massive team of different departments working in tandem towards a common goal. Regardless of the size of the team, the principles are the same: there is a common goal, operations, organization, and communication. When a team works well, these feel like a well-oiled machine. This takes a bit of social intelligence to activate the right people for the right job in the right way. Every team could benefit from taking the VIA survey to learn about each member's strengths and to maximize everyone's impact toward the goal.

MUSIC PROFESSIONS

Rehearsal Studios

1. *noun*: a music rehearsal space is a room or number of rooms dedicated to music-making. A professionally sound-proofed practice room is an acoustic environment defined by its purpose and layout, designed to keep sound inside and unwanted sound out. Spaces can combine rehearsal and recording functions.

Placing Professions into Studios

There are so many ways individuals work in the field of music. The studio chart below categorizes professions into four main groups: performance, creation, education/research, and industry. These professions have common activities, circumstances, and/or required skills. For instance, the group "LIVE PERFORMANCE" consists of people who:

1. Perform music in front of an audience
2. Do so in a live setting
3. Require instrumental or vocal skills/training
4. Memorize, read, or create their parts
5. May contend with elements of performance anxiety
6. May enjoy the "spotlight"

These individuals as a group may exhibit similar traits and needs. They may find their strengths have called them to their profession or that they synergistically serve and engage them in their role. It is important to recognize that professions can cross studio lines. An independent musical artist is also an entrepreneur; often doing equal work on both business and creative tasks. For this reason, these categorizations are not absolute but aim to group the professions in the most logical way.

Ways to Use This Section

As you hear examples of people in musical fields empowered and fueled by their character strengths, you may be inspired to cultivate, use, or grow certain of your own. You can use the First Set as a resource for that.

You may recognize yourself in the profiles of music professionals. Do their stories help you reaffirm what you are already doing with your strengths? Or do they inspire you to infuse your actions with different signature strengths?

On the contrary, you may read about someone in your exact job and see how they do it from a totally different perspective. Two people going after the same ends can employ very different means. This is the power of leveraging your individuality. And it is what will make you successful and unique. With self-awareness, you will see how easy it is to rely on your strengths to take you through tasks and help you cultivate other strengths. Your signature strengths are superpowers that can be called upon at will.

You may also be "blind" to a strength of yours because it's so "natural" to you. For instance, someone who has "love" as a top strength might just assume all people value personal and professional relationships and take extra care to nurture them. Besides, it's super easy (to you). It is important to realize the things we do joyfully and in flow come from our strengths. Someone who does not value love and relationships will have different business values and actions.

Another way to use the following chapters is to seek professions that may align with your strengths profile – jobs that you may be "naturally strong" at. When you have a good match between what you do and your signature strengths you are more engaged, more creative and experience greater well-being.

Placing Professions in Virtue Categories

Also, it is important to know that professions have multiple layers, just as people do. In the same way, professions can be placed in multiple categories, depending on what you focus on. For example, a teacher may mostly live in the wisdom category with its focus on learning, but can you also see how courage, humanity, and justice all factor into that role? Any teacher can tell you that a large portion of their time is dedicated to many aspects outside of simply "teaching" (classroom discipline, lesson planning, administrative work, etc.).

It is good to remember that professions cited as examples of these virtue categories are *not exclusive* to that category. Each job calls for a myriad of different strengths and skill sets. Instead, the chosen professions cited most strongly express the *essence* of that virtue category. The profession itself is analyzed for which strengths it most exemplifies.

A successful person in that role may have most of their high strengths in that category. And then again, they may not but have other strengths that bring a fresh approach and uniqueness into the modus operandi of their profession. You will see that tends to be the case for many of the profiles you will read. Perhaps that is an implication of being in a creative field – practitioners may already lean a bit "left of center."

There are so many professions in or tangential to music. If you are reading this, you are most likely in or studying to be one of them. Let's look at the variety of music professions (Table 3.3–3.6).

THE REHEARSAL STUDIOS

Table 3.3 Studio A: Music Performance Professions

Studio A: Performance
Live performance: Artist/Performer Accompanist Backing Musician DJ Conductor: Band, Orchestra, Choir Church/Worship Musician Director: Opera, Musical Theater, Ensemble Touring Musician
Recording: Session Musician Engineer: Recording, Sound, Mixing, Mastering Vocal Engineer Producer

Table 3.4 Studio B: Music Creation Professions

Studio B: Creation
Music creation: Composer Songwriter Producer Film Scorer Video Game Composer Lyricist Music Arranger Transcription/Notation Sound Editor
Media creation: Artistic/Creative Director Music Videographer Music Video Director Music Marketer

Table 3.5 Studio C: Music Education and Research Professions

Studio C: Education/Research
Education: Music Teacher (PreK-12) College Music Professor Private Lesson Instructor Music Administration/Leadership Performance Coach Music Tutor Choreographer Dance Teacher Mentor
Research & writing: Music Critic Music Librarian Music Theorist Musicologist Music Researcher Music/Holistic Wellness

Table 3.6 Studio D: Music Industry and Specialty Professions

Studio D: Industry/specialty
Business Roles: Manager: Talent, Touring, Day-To-Day, Digital, Business, Brand/Sponsorships Entertainment Attorney Agent: Talent, Booking Entrepreneur, Tour Promoter Music Supervisor, Music Publisher Record Label: Marketing, Touring, A&R, President, Executive Roles Music Technology
Marketing/branding roles: Publicist Music Photographer Social Media Platform/Design Specialist Playlist Curator Branding Agent Marketing Specialist
Merchandise and repair: Music Merchandise Sales Instrument Maker App or Platform Designer Instrumental Repair or Maintenance Piano Tuner
Specialists: Music Therapist Expert Witness Musician Stylist Lighting Director Stage Costume Designer

VIRTUE CATEGORIES
+
MUSIC PROFESSIONS

A Look at the Virtue Categories

The 24 character strengths are classified into six virtue categories: courage, temperance, wisdom, transcendence, humanity, and justice as seen in Table 3.7.

Table 3.7 Virtue Categories with Strengths

Virtue Category	Strengths
Courage	Zest • Honesty • Perseverance • Bravery
Temperance	Self-Regulation • Humility • Forgiveness • Prudence
Wisdom	Love of Learning • Judgment • Perspective • Creativity • Curiosity
Transcendence	Spirituality • Gratitude • Hope • Humor • Appreciation of Beauty & Excellence
Humanity	Love • Kindness • Social Intelligence
Justice	Fairness • Leadership • Teamwork

How Are Professions Placed into Virtue Categories?

Professions are not exclusive to any one virtue category. However, one can place them in the categories by making informed assumptions about the type of skills, attitudes and aptitudes required for each. The named professions chosen for this chapter epitomize and strongly express the essence of that virtue category.

Job postings often directly ask for strengths as qualifications. One posting for a streaming platform music sales position wants someone with, "a true love for [platform] and a passion to produce creative and innovative solutions for our key clients and partners. You should be confident while staying humble, work hard and know how to have fun!" Can you spot the strengths named in this job posting?

Each profession calls for a myriad of different strengths. A successful person in that role *may* have signature/high strengths in that category. Or they may have a variety of strengths *across* the categories. Notice in the following case studies that actual practitioners often possess "a rainbow" of signature strengths; strengths from more than three categories (some possess four or five). You might even perceive these people as being "well-rounded" as their character strengths are in multiple categories.

For example, we would imagine a successful entertainment attorney would exhibit many strengths in the JUSTICE category. However, the attorney we interviewed didn't have any justice strengths in his signature top five! This is an example of leveraging individuality. Doing a job in your own, fresh way can make you stand out amongst your peers in the same profession. You can form your own "brand" or modus operandi based on your natural signature strengths.

Another way to use the following chapters would be to seek professions that may align with your strengths profile – jobs you may be "naturally strong" at as they speak to your top strengths.

Methodology of the Interviews

All participants first completed the VIA strengths survey. From there, they had the choice to fill out a detailed intake form explaining ways their strengths show up in their work/life or to complete an in-person or zoom interview. Those who participated in the interview gave an additional personal or professional best self story (Keltner, 2022) explaining a time in which they felt "at their best," firing on all cylinders, and totally in flow with what they were doing.

Part of the initial thrust for writing this book was to highlight and celebrate the proprietary mix of strength spices that make you a unique flavor in your artistry, industry, and life. Sometimes, we are encouraged to change, fit a mold, and become someone we are not to "succeed." However, the happiest you will be in your art, your job, or your life is rooted first in proudly being yourself. We are all so multivalent and layered. Bringing the variety of these seeds into the fertility of the arts creates gorgeous beds of flowers. Think of the icons in music and the ways they brought their true selves into the establishment. There is usually an element of bringing something so personal and new into the scene and therefore standing out while creating something authentic and enduring. If you feel disconnected from your own truth, what can you really say through your art? Solidly connecting your actions, words, and efforts to the core of who you are artistically and professionally is a key feature of being in harmony within yourself and in your external world. Give yourself permission to be you!

The interviews below are a great example of the way practitioners in the field achieve success by infusing the demands of their job with their individuality. As you will see, the interviewees bring their many facets to the work they do and leverage their strengths to shine!

THE COURAGE PROFESSIONS
Zest • Honesty • Perseverance • Bravery

People who have high strengths in courage may gravitate towards professions that are more front-facing and exhibit "higher pressure." They may have to bravely face critics, audiences, or investors. They exhibit perseverance by working tirelessly with clients or students or learning difficult music. They speak their truth through lyrics or find truths through research. They approach their day and tasks with zest! Here are some professions that engage the COURAGE strengths (Table 3.8).

Table 3.8 Courage Professions

Studio	Example Professions
A: Performance	Artist/Performer Director: Opera, Musical Theater, Ensemble
B: Creation	Songwriter Artistic/Creative Director
C: Education & Research	Private Lesson Instructor Musicologist
D: Industry	Entrepreneur Publicist Manager

Courageous Professionals

Brian Wilkins

MANAGER • MUSIC TECH ENTREPRENEUR
Bravery • Kindness • Humor • Judgment • Curiosity

Best-Self Story When he was working in radio, Brian discovered he wanted to work on the music industry/label side rather than continue the broadcast media side. A friend of his moved to New York to work for Sony Music and had intel on a radio promo assistant position that was opening. Brian would have been a great fit. He got an interview for this position, but, unfortunately, it coincided with his annual on-air radiothon hosting duties for the Make-A-Wish foundation. He requested a different interview day, but they basically told him he was lucky to have this spot and there were many people lined up for this. He decided it was more important to stay for the cause and if he had to forfeit the job, so be it. He didn't get that job, but the radio station where Brian was working set a record for raising money for Make-A-Wish. Shortly after, he ended up getting a better position within SONY music. At the time, those jobs were few and far between, especially for someone out of state and who didn't have direct label experience. It was tough to turn down that type of opportunity, but he had to make the choice that aligned with his personal values.

Strength Spotting Bravery was at work in Brian's story. It took courage to make the decision to forego that opportunity in favor of his own convictions, values, and obligations. His kindness shows in the way he honored his duty to a charity event over an exciting job opportunity. And clearly, judgment was also active in his story in the way that he chose his personal commitments and values.

Signature Strengths

Bravery (Courage)

"Professionally as a manager, you have to say some contrary things sometimes and you must be real with the people with whom you're working. That's not always easy. Telling people, "no, that's not what you should do," or "no, I don't like the choices you made in this song" is essential because your livelihood depends on it." In his personal life, Brian has made many decisions that brought him out of his comfort zone. Growing up in the rural south and moving to New York City was a huge act of bravery.

Kindness (Humanity)

This shows up for Brian as appreciation for things and the efforts people put into experiences and relationships. Many times, managers must deliver bad news or voice an opinion that may be contrary to what the artist wants to hear. He has learned through experience different ways to say difficult messages with kindness. And notes it's often a rough, blunt terrain in the music business. "It's often the opposite of kindness in my industry. But it doesn't do any good to tear others down. There's a difference between speaking your mind and being rude and unsupportive."

Humor (Transcendence)

Everything Brian does includes humor. There is a relaxing, disarming, communal spirit to Brian's humor that brings people together and helps people see each other's point of view. It's a strength he uses naturally to find a common ground in relationships. Brian enjoys the lightness that humor brings.

Judgment (Wisdom)

Brian says he's a textbook example of a Gemini who is often weighing decisions – small, medium, and large. It is very natural for him to analyze different sides of arguments or decisions. Seeing the pros and cons of every choice comes naturally to him.

Curiosity (Wisdom)

Brian loves deeply learning about everything he's interested in. He is encyclopedic about the seminal people, history, and landmark events in modern music. But apart from that, he takes deep dives into other subjects, especially podcasts and videos describing how things are built or created – from businesses to boots! He loves authentic and true entrepreneur stories and he's always curious about how people did what they did and how to apply principles to his own endeavors.

Celeste Tauchar
(Talker)

ARTIST • SOCIAL MEDIA MANAGER

**Love • Humor • Curiosity • Appreciation of Beauty
& Excellence • Creativity**

Best-Self Story Celeste told the story of working on her EP (extended play – refers to an album consisting of anywhere from three to four songs, shorter than a full long play or LP album) during the pandemic. Her good friend and producer moved farther north into a house with his wife during the pandemic. To assure everyone's safety, Celeste got COVID tested and fully blocked out a month-long period to work together on her EP.

This month turned out to be her favorite musical experience because she was able to be so dedicated and focused. In the mornings, she ran her social media management business and attended online meetings. Afternoons were completely blocked out to work on the record. Celeste felt fully present in her creative headspace and was able to dive in with little distraction.

Strength Spotting A lot of Celeste's signature strengths were working for her on this project. She exhibited a love of two different sides of her personality/brain – the businesswoman and the creative. She was able to nurture both by allowing them to exist in their own space. She appreciated the beauty of the music she was making and felt in flow with the creative process. In choosing her friend/producer to shelter in place with, she had excellence all around her which motivated and supported her project. Her decisions had set her up for success.

Signature Strengths

Love (Humanity)

This is a big strength for Celeste and universally touches all she does. Celeste is very relationship driven – whether with music or friendships – she always seeks community. Part of this goes with the territory of pursuing a music career but, most importantly, she cares most about being a part of a larger music scene/community and having friends with whom she can collaborate. Celeste holds sacred space for friends. For example, she and her roommate/best friend dedicate time to "sparking joy" around the apartment. This practice, inspired by professional space organizer Marie Kondo, focuses on keeping objects which give you joy and letting go of others that do not. Love is intrinsically valuable to Celeste, and she directly correlates her strongest friendships with some of her best professional opportunities. Whether it is building her friend base, her fanbase, or her personal connections, everything she does is love!

Humor (Transcendence)

Just like love, humor is innately part of Celeste's personality. Humor is a vital part of her friendships, onstage persona, and social media personality. She explains that humor is not just about being "haha" funny, but also a general lighthearted approach one brings. She feels most things aren't that serious at the end of the day. Things can feel heavy and life-or-death but it's music – it needs to be fun! Celeste reminds herself to keep it light.

Curiosity (Wisdom)

Lately, Celeste has been thinking a lot about the concept of specialization – leaning into one thing to become the best at it. She understands she could practice the guitar for hours a day, but she questions the importance of that. Instead, she prefers being good at several things and feels if she's so focused on one thing, she won't be a well-rounded human. "It's important to be good at your craft but if you're not living, what are you even creating for?" She makes traveling a priority so she can see and experience things. She even took a virtual Spanish class and is always trying to develop other interests as she feels this is ultimately a benefit. Her curiosity follows things she finds beautiful or interesting, and this fuels her creativity. Her strengths interconnect and drive each other.

Appreciation of Beauty and Excellence (Transcendence)

Celeste finds the definition of this strength interesting – especially the part about appreciating excellence. She gets emotional watching excellence and finds it stimulating to put it in this perspective. Celeste loves a good story of someone having a dream, working hard, and executing it. "You can frame that as loving ambition or seeing yourself in that person – there are elements of both – but I truly appreciate universal human desire and striving."

Her pursuit of beauty pairs with her curiosity. She camps a lot, loves the outdoors, and tries to get herself into beautiful spaces to detach from everything. "Being in beauty makes you want to be better." She consciously projects the way she sees herself – as a creative, artistic, empathetic individual. She feels disconnected from that aspirational version of her best self if she's in a bad space.

Creativity (Wisdom)

Celeste exudes creativity in her work as an artist but also in everyday life. She's a generally creative person who is always looking for new ways to make things special. For example, she and her roommate are always trying to find creative ways to "spark joy" or elevate a moment. Recently, they decided to have a fall-themed day – getting pumpkin-spiced lattes and buying fall-scented candles. Although this may be a subtle example, she is always trying to come up with ways to make life more beautiful and fun.

THE TEMPERANCE PROFESSIONS

Self-Regulation • Humility • Forgiveness • Prudence

Those who have strong temperance have the stamina for the types of careers requiring patience and exactness. Engineers work tirelessly for weeks on minutes of music exhibiting tons of self-regulation. Music publishers pour through submissions exhibiting great prudence in their selections. Worship musicians use music as a tool of humility and forgiveness to serve a greater purpose. The act of studying music theory or learning an instrument takes a huge amount of self-regulation through years of practice and education. Here are some professions that engage the TEMPERANCE strengths (Table 3.9).

Table 3.9 Temperance Professions

Studio	Example Professions
A: Performance	Church/Worship Musician Touring professional Engineer: Recording, Sound, Mixing, Mastering
B: Creation	Music Arranger Transcriber/Notation Sound Editor
C: Education & Research	Music Tutor Music Theorist Professor
D: Industry	Music Publisher

Temperate Professionals
Martin Blessinger

COLLEGE PROFESSOR • COMPOSER • MUSIC THEORIST

Love • Forgiveness • Humor • Love of Learning • Gratitude • Prudence

Best-Self Story Martin's best-self story encompassed the whole process leading up to the performance of one of his compositions for piano and cello. He really enjoyed every aspect of composing this piece. He came home each day from work, truly energized to dive in and write, which was not the case for all pieces. As a composer, sometimes it's a struggle and sometimes it comes easily. Most creatives can relate to this sentiment. This time felt not only easy, but exciting to him. He got lost for hours, flowing with inspiration, during the totality of the composition process.

The husband-and-wife cello/piano duo whom he had chosen to debut his new piece poured themselves artistically into learning it. They took it seriously and put a lot of love into the details. The duo made sure they were not only expressing his vision but inserting their own voices into the phrasing and expression.

The night of the premier was probably the best musical performance he's ever had of one of his compositions. Everything fell into place exactly as he had hoped. The performers were met with a full house that embraced the piece and their playing. The audience was so eager to clap after movements, that any pause from the duo would sometimes result in spontaneous early applause from the audience. Martin had written musical jokes into the piece and the audience ate it up, laughing at each funny moment. Everything went extremely well and felt aligned.

Strength Spotting Martin's best-self story exhibited a lot of his top strength: love. Love guided the process. This piece felt different because he loved coming home to immerse himself in the creativity of composition, even after long days at work. This was literally his labor of love.

The love continued with his fondness for the duo and the love they shared for each other. As a married pair, they intimately knew the other's playing and were able to use that spark to bring life to the work. Martin's enthusiasm for rehearsing with them was also part of this love relationship.

Finally, the love he received from the appreciative audience made this moment perfect for him. Humor, his third strength, showed strongly in this piece. It's not something most composers typically consider when writing but for Martin, this signature of his was important to include. He really loved that the audience "got" the musical jokes. His gratitude was so evident, it seemed to beam from him, as he reverently told the story of this performance. His whole demeanor lit up and conveyed his thankfulness.

Signature Strengths

Love (Humanity)

Love is probably Martin's most cherished strength. He was not surprised that this was his number-one character strength. Love helps him feel healthy and connected to something bigger than himself. He feels the importance of love in his relationships, family, work, and interactions – with people and creatures big and small. Love abounds for Martin! As his number-one tool, he leans on it in times of trouble and when things get difficult. When asked what it would feel like to him in a world, job, or relationship where he was not able to give or receive love, he said it would be stifling, unsafe, isolated and he would feel "boxed in."

Forgiveness (Temperance)

In Martin's stories, love and forgiveness really do hold hands. He discussed his semi-recent divorce and how forgiveness became one of his most cherished tools. He learned how important it is because it allows you to let go and move on. As an act of love, it also releases the other person to move on. He said even if there is no apology, you're still leaving the baggage behind that is a net positive. It's not only an act of love but also an act of self-love that releases you from an unhealthy situation.

Humor (Transcendence)

As we have already learned, Martin really values humor in all aspects of life: personal, professional, and even musical. He processes the world through humor. Martin describes finding humor in his relationships as "all about creativity. It's about holding something back with the right timing, then releasing it and seeing what the other person's brain comes up with." He likened this to playing tennis with someone else who is also good. All his closest relationships fall into this category – he likes to play with pros. If there's no humor, there's no spark for him – "it's kind of dead." As a music theory teacher, he uses humor to keep students on their toes, "They don't expect music theory to be funny or entertaining and it keeps them really engaged in the lessons."

Love of Learning (Wisdom)
As a professor, a love of learning is high on Martin's strength list. Not only did he undergo massive amounts of education – but he has also never been off a school schedule since kindergarten – he is committed to a lifetime of learning. When he is not dishing out knowledge, he is taking it in, engaging in all sorts of expansion and enrichment. He goes to conferences, stays current in his field, reads all types of things about space, philosophy, and cosmology, devours documentaries, learns French and Spanish, and takes Zoom German conversation lessons. When asked what his life would feel like if he couldn't learn, he responded, "I literally wouldn't know what to do with my time."

Gratitude (Transcendence)
Discussing gratitude filled Martin with joy. He makes sure he always appreciates those who have helped him – every night lying in bed to end his day. Martin expresses his gratitude through a polite demeanor and by acknowledging how relatively easy his path has been in comparison to others. He counts his blessings and actively keeps this awareness in the forefront of his mind. Two of his highest strengths – fairness and prudence – tie into his gratitude. He says these all show up for him in teaching situations where he is cognizant and empathic to the different paths and struggles of people in the classroom. He makes sure to deliver his lessons to all his students in an accessible, respectful, and patient way.

Prudence (Temperance)
Even though this strength is sixth, Martin shared that prudence shows up for him in all his dealings and relationships, especially in written correspondences. It's a very important aspect of his personality. He scans every text and email for meaning and thinks deeply about the best way to respond to be precise and avoid inflammatory language. Others seek him out as an "emotional proofreader" when they are sending difficult or important emails or texts. His prudence is a super-strength! He compared this with his bottom-ranking strength, bravery. In the bold definition of bravery, prudence seems to be the perfect counterpoint. He would not feel strong acting "bravely" without really analyzing a situation. He would rather, very level-headedly, assess all factors and make careful choices.

Martin self-reports that his signature strengths felt like a true snapshot of him. He said it was powerful to see them and relate them back to the way he approaches his work and creativity from a more internal place. "It's so easy to get lost in musical careers because everything gets filtered through only talent or results."

Jean Caze
TOURING/PERFORMING MUSICIAN
Appreciation of Beauty and Excellence • Gratitude • Love
• Hope • Prudence

Best-Self Story Jean remembers one of his first performances at age nine. His peers each stood up and played part of a melody. He stood to play his trumpet and afterward the crowd erupted into applause – which he did not expect! It wasn't a particularly difficult solo that he had rehearsed to the maximum; he was just sharing his joy and passion for music, which clearly came through in the way he played. It was one of those natural gift moments that he continually revisits to remember why he does what he does and how you can really touch people with your art. In other performances, he has felt like he was not even playing, like he was being "used" as an instrument and someone else was playing through him. Also, he remarks on the way people look at you after those moments, like they've seen a ghost.

Strength Spotting In Jean, appreciation of beauty and excellence is operating in two ways – he is receiving the gift of music and actively creating beauty and joy. He also appreciates the beauty he finds in others. His gratitude is far-reaching and abundant. He is grateful for his talent, and even for the way he feels he has "channeled" music (feeling like you are being used as an instrument of a higher power to express music). In this way, it acts like a true gift. One must love what they are doing to commit to a livelihood in music. If you are reading this, you know it is often assembling career puzzle pieces with no clearly divined path. There are many more predictable careers, so it takes a lot of love to sustain stamina and find new fresh ways into music. Also, to get to where Jean is in his career, he had to exhibit the "golden mean" strength of making prudent choices to best navigate and operate in many different musical systems and settings.

Signature Strengths

Appreciation of Beauty and Excellence (Transcendence)
Jean sees beauty all around him and appreciates how it comes into his life. Whether it's scrolling on Instagram, looking out the window and noticing the sky, appreciating beauty in a person, noticing symmetry, or listening to elegantly simple music, he gravitates to beauty. As a Libra, he is always looking for beauty in balance, even mechanistically when he is tapping a finger pattern or playing a melodic line, he is very aware of the beauty of symmetry in all things.

Gratitude (Transcendence)

Jean is grateful to be able to learn, grow, and forgive himself for his mistakes. He is grateful for music and how he can support himself and his family through his dedication and work. He is thankful for good health, the people he has in his life, and all of life's necessities – a place to sleep, food to eat, and for not living in a war-torn country. He is grateful for his healthy perspective and his ability to not overthink things and give himself rest. Clearly, Jean's gratitude is an anchoring strength that he has in abundance. Gratitude itself is a powerhouse strength that has far-reaching benefits.

Love (Humanity)

Love and affection are so important and grounding to Jean. When he's on tour, he's away from the people he loves and that's difficult for him. He makes sure to connect through zoom and phone calls but also recognizes that he feels better when he's around the people he loves the most. Being around his family and loved ones makes him truly feel complete. Jean knows it is also important for him to learn different ways to love the different people in his life. "Everyone gives support in different ways, and you need to learn how to see and take this in. When love is a big piece of the equation, you will do the work to fulfill their needs."

Hope (Transcendence)

He believes life is what you make it and you have the power to change your life with every minute and activity throughout the day. Jean has seen many examples in his life where he started in one place and ended up somewhere else due to faith and the hope to keep going. This also includes putting in the effort and seeking out the right type of help and information. Although there are things in the world he hates and yearns to change, he empowers himself by fixing things he can change within himself. He hopes that what he does makes a small ripple effect to help others get to a better place and shift and interrupt negative cycles.

Prudence (Temperance)

Regret poisons Jean, so he tries to be very careful with his actions and words. He knows people have great memories for painful experiences and he doesn't want to cause anyone harm for any reason. "When people die, you remember how they made you feel." Jean's careful choices have built him an incredible reputation that commends him through all of his professional endeavors.

THE WISDOM PROFESSIONS
Love of Learning • Judgment • Perspective • Creativity • Curiosity

You may know individuals with a lot of wisdom strengths. They are the ones who seem to always be expanding their knowledge base. These are the people who pursue their field to the highest degree and then some. Creators who explore genres outside of their own training, culture vultures who consume new music for breakfast, or researchers who travel far and wide into exotic archives to discover. Here are some professions that engage the WISDOM strengths (Table 3.10).

Table 3.10 Wisdom Professions

Studio	Example Professions
A: Performance	Accompanist/Backing Musician
B: Creation	Film Scorer Video Game Composer
C: Education & Research	College Music Professor Music Researcher Mentor
D: Industry	Music Supervisor Playlist Curator Branding Agent Branding/Marketing Specialist

Wise Professionals
Cassandra Claude
ENTREPRENEUR • COLLEGE PROFESSOR • MENTOR
Love • Spirituality • Love of Learning • Social Intelligence • Creativity

Best-Self Story As a powerful mentor and teacher, Cassandra feels at her best when she is creating a platform for others to be their best. She spoke of a time when she held a massive leadership conference within her own company. She had been doing smaller gatherings throughout the year but this one was a huge, culminating event. Cassandra remembers feeling consciously connected to her core and aligned with her life's purpose during this conference – she was meant to do this. She greatly enjoyed bringing industry experts and interesting speakers who brought different perspectives to the musician audience. This is her specialty – tying in inspirational figures from various disciplines to speak on a common theme. This has proved effective in her leadership, teaching, and other business pursuits. This is also part of how she approaches her own goals, by bringing in perspectives; far-reaching practices; and other cultures that teach, expand, and excite whatever audience she serves.

Strength Spotting Cassandra's social intelligence played a huge hand in organizing this conference. Her ability to bring in great speakers comes from the vast social networks she maintains. She is excellent at spotting and deploying talent in unexpected ways. Love of learning is the core of her story. Though Cassandra brings educational experiences to her clients and students, she gets just as much out of it for herself. Love and spirituality infuse the types of events she holds. Cassandra always hires a speaker who specifically focuses on self-love and loving one's craft. This is noteworthy as industry events rarely focus on these types of internal elements in favor of more extrinsic topics like networking, financial success, metrics, etc. Cassandra makes sure that her seminars are grounded in a deep respect for each musician's gifts. She helps them find true meaning and purpose. She has an online class dedicated to coaching musicians to find their "why." Her creativity was an overriding theme for the whole concept and design of this event.

Signature Strengths

Love (Humanity)
It is a wonderful thing to really love what you do. For Cassandra, this can sometimes present challenges as her abundance of joy and energy for her work can take up personal real estate designed for loved ones. Cassandra's deep-rooted love for what she does gives her great purpose. She also feels strong love for her partner, family, and friends. Love is a throughline in her life and guides many of her decisions and helps her keep things balanced.

Spirituality (Transcendence)
Cassandra has a wide view of spirituality that incorporates God, intention, and energy. She deeply ponders her life's purposeful journey. Embracing the spiritual power of thoughts and words, she has manifested many amazing people and experiences. She is in awe of divine timing and serendipity. She keeps sacred space for her spirituality in all she does. This shows up in the way her current relationship came into her life and the many other blessings for which she is actively grateful. Cassandra also helps her students spiritually connect to the music they are performing. She does this in a very reciprocal way – directing them to both bring themselves into the music and to let the music enter them. Her students cherish the sacred space she keeps for "connection" in the teaching studio.

Love of Learning (Wisdom)
Cassandra is a lifelong learner. She systematically goes through all topics of interest and throws herself fully into the pursuit of knowledge. She desires to truly master new subjects and has many certifications to show for it! It's not enough for her to seek the knowledge; she aims to master it through formal study. Currently, she is interested in pushing herself physically by conquering more advanced machinery at the gym. Her love of learning is equal parts intellectual, spiritual, emotional, and physical. She feels she is in competition with herself, but not with others. It means more to her own sense of accomplishment to learn and achieve something challenging.

Social Intelligence (Humanity)

This strength is an important tool for teachers and Cassandra has it in spades. She has always possessed an empathic, innate sense of others and their emotions and needs; especially related to teaching music. However, in her recent years, her interpretation of social intelligence has grown to include not just reading different situations but knowing how to respond. She is learning in her current relationship how to adjust to her partner's needs, and this is operating as a tool in other parts of her life; learning to better meet the needs of others around her. This strength is very much on a growth track that connects to her love of learning. She continually works on interpersonal relationships to deeply understand others.

Creative (Wisdom)

Funny enough, Cassandra at first didn't feel like she was creative. Clearly, however, her approach to many of the things she does comes from a different perspective than expected. For instance, the way she runs her teaching studio, filled with both physical, emotional, and spiritual work, is quite different from most voice teachers. For this distinction, she is sought out by students, the press, and other educational institutions. Her students have a unique experience under her direction. Not many students can say that their teacher wrote a book on the chakras' connection to the singing voice! This is yet another example of her creative, out-of-the-box way of thinking about voice and performance. And a great example of strength blindness – we often don't notice and take for granted our top strengths because they are just the natural way we think or operate.

Christopher Mazur

MUSIC SUPERVISOR
Curiosity • Bravery • Kindness • Zest • Love

Best-Self Story As a music supervisor, Chris says his super-power is problem solving. He is at his best when things teeter on the impossible and unsolvable. His best-self story was about an early project in his career working for an advertising agency. It was for a major client using the power of music to showcase the power of data. He chose to do something with algorithms that hadn't been done before. The agency hired a famous musician who liked working in this space. Together, they developed a plan to use live U.S. Open tennis tournament data to generate live streamed music. Thus, began six months of trying to figure out this problem in the studio with this musician and his engineer. They began curating "musical sounds" so the result would be listenable. At some point, a secondary company was hired to help build out the algorithm.

Soon they realized the algorithm, built by a mathematician, functioned on decimals. The music was built on fractions and was not compatible. They kept hitting the same wall and the project came crashing to a halt. They were down to the wire, two months out from the tennis tournament, and no one knew what to do. The music sounded "like trash" and the mathematical problems kept happening.

Determined, Chris hired the "smartest people he knew" and together, they reverse-engineered the algorithm and solved the problem. Chris credits his success to leveraging the talents of others and keeping a positive "yes, I'll help figure out this problem" attitude in the face of dead-ends and issues. He also notes that he is unphased by fame. Chris explained, "At some juncture in this line of work you will be faced with large numbers in a negotiation, or you will be having a conversation with someone who is incredibly important or famous. If you allow being star-struck to bleed into that, you won't get the best results. You need to be able to engage and act with some humility." He asserts that the combination of problem-solving and being unphased by fame has repeated itself over time in all his most successful projects. Correspondingly, his best projects were the ones that presented the messiest and most chaotic challenges.

Strength Spotting Listening to Chris passionately describe this work scene, one can feel how each of his strengths energizes each other. His curiosity and bravery are "all in" when attempting to do something no one else has yet done. It takes resilience to dust yourself off and try again after something is failing for four months. That kind of perseverance served him and his ability to stay cool under pressure and maintain control. Zest returned daily with his refusal to give up. "No, we're going to fix this," was his mantra throughout the project. Zest acted like the "battery pack" to his other strengths. You can feel the passion, love, and pride he had for this project as he spoke. Solving difficult problems can be addictive. Chris has made a reputation for himself logging multiple successes in challenging work settings. He appreciates jobs that open his mind to new solutions or possibilities and make him grow.

Signature Strengths

Curiosity (Wisdom)

Chris was surprised and pleased that this was his number one strength. He had specifically been challenged at work to cultivate and grow his curiosity. He spent the better part of the last three years trying to grow this strength by asking better, more engaging questions that go beyond the surface and don't assume a baseline answer. He practiced being very open-minded, removing assumptions, and allowing space for the other person. He used curiosity to be more imaginative to get through obstacles. "If you are really held within a strict 'this is my logic' box, then it's unlikely you're going to be able to solve problems. Curiosity allows you to acknowledge that these logic constraints are a function of repeated emotion over time. Solutions can go beyond what you have done repeatedly to solve that problem." Chris' curiosity fuels his scientific mind, allowing him to go into situations without preconceived notions.

Bravery (Courage)

Funny enough, Chris said he felt like bravery is one of his biggest weaknesses because it gets him in the most trouble. All our strengths can be used in a balanced way (the golden mean), or we can over and/or underuse them. When people describe their strength as a weakness, most likely they are feeling it is being overused. In Chris' case, when he feels like something is wrong, he is incapable of not addressing it. He feels impulsiveness to act. He said in these moments, he is so confident in himself and cares so little about what the other person might think of him he will fight the injustice strongly even if it means someone will think less of him.

Kindness (Humanity)

Kindness for Chris intertwines with bravery and compels him to act. He feels kindness towards people when they are being overlooked or taken advantage of. He says the "justice warrior" in him kicks in and he feels very intense empathy. This directly ties into his bravery.

Zest (Courage)

Chris' zest materializes as a self-described addictive personality. When he enjoys something, he tends to overdo it. He has hobbies he started 18 years ago that are still great passions – from sports to board games, he will "nerd out" and get super into them. He simply loves unpacking new sets or systems of rules. He feels these check all the mental fitness boxes he needs. "It can be the silliest thing, from immersive or world-building games to a checkerboard you fill out with colors. There's something about trying to dissect a new set of rules to get the most out of them that I really enjoy. It's my mental gym." Also, he notes if something doesn't interest him, he's low-key and disengaged.

Love (Humanity)

At first, Chris was surprised that love is in his top five – it made no sense to him. However, it made a lot of sense when he listened to his passion for everything as he described his life. He loves everything he does. As a matter of fact, there's not one thing he actively makes a part of his life that he doesn't love. In terms of the other side of love, valuing people and relationships, he said he has really learned how to love in the past decade, and he has actively worked on this, pushing through issues he had to get through. The work he has poured into this is showing up in his top five strengths! The act of working on love itself signifies its huge value to Chris.

THE TRANSCENDENCE PROFESSIONS

Spirituality • Gratitude • Hope • Humor • Appreciation
of Beauty & Excellence

Those high in transcendent strengths understand the importance of the feel, meaning, vibe, and look of things. They are the producers and composers who create stunning aural worlds, the music tech people who value design as much as function and the music video directors whose vision becomes their moniker. They are the performance coaches who understand the value of humor to offset the ardor of work. And they bring a higher sense of purpose and spirituality to all they do. Here are some professions that engage the TRANSCENDENCE strengths (Table 3.11).

Table 3.11 Transcendence Professions

Studio	Example Professions
A: Performance	DJ Producer
B: Creation	Composer Music Videographer Music Video Director
C: Education & Research	Performance Coach Music Critic Music Wellness Fields
D: Industry	Record Label: Promotion, Marketing, Touring, A&R, President, Executive Roles Music Technology

Transcendent Professionals
Daniel Loumpouridis
PRODUCER
Humor • Appreciation of Beauty & Excellence • Leadership • Hope • Judgment

Best-Self Story Daniel said that he can recall entire days when he felt totally in tune and flow with the universe. Those days felt like him at his best. One example was the first time he worked in a legitimate recording studio as a producer. He was flown out to Nashville to work for an artist and felt he was entirely unprepared. He had never worked in a real studio like this before. There was an engineer on the computer and hired musicians playing the instruments. This was the first time he was not responsible for recording and/or playing something and could truly just "be the producer." He feared he wouldn't have a place in this process but when he arrived, he found his role immediately. All these people were in the room just waiting for the captain. Daniel didn't realize he was going to be "the captain" until he walked into the room. It was such a natural fit and he was "absolutely crushing it." He listened to the song and then immediately began giving all of the different musicians instructions. "I was able to shape these people and this experience and bring everyone together in a community and really get the room working well."

Daniel was recently reflecting on this experience with his studio partner at their Hollywood recording studio. "A producer's number one job is vibe curation and to make the room feel good. Everyone in the room should be comfortable and everyone should be inspired. If I can make that happen, we're golden." Upon further reflection, he said all his best moments were bringing a room of people together to do something they didn't know they were going to do. The vibe creation emerges from guided co-creation.

Strength Spotting Daniel clearly showed leadership in this story. The way he assumed this new role showed how his natural abilities just needed the right setting to shine. He orchestrated a room of people and leveraged their talents to make a cohesive musical product. Not without mention, they had to trust him to accomplish this. Knowing Daniel, he undoubtedly used humor as a tool to make people comfortable and gain that trust. His appreciation of beauty and excellence was core to the whole endeavor. As he told details of what he had the musicians try and do, he lit up with excitement. Curiosity was apparent in his experimentation with things outside of his own personal capacity. His imagination could run wild in this project in a way that couldn't happen with him doing everything by himself. What a delicious recipe when leadership combines with good teamwork. This freed Daniel up to be more expansive and creative.

Signature Strengths

Humor (Transcendence)
Daniel says everything's funny. "Not in the existentialist, Camus way, but it's not NOT that. The idea of life being a cosmic joke is cynical but if you ever take it too seriously, you're messing up. People who know me best know everything I say has 10% humor in it. Nothing is ever the end of the world – except maybe climate change – and when you keep that in perspective you connect with people and don't take things too personally. Everything has a lightness to it. It's a coping mechanism to keep things from feeling so serious all the time." He uses humor and lightness to access greatness from artists.

Appreciation of Beauty and Excellence (Transcendence)
This strength is the same for him as humor – it's a daily practice. "Good art and appreciating beauty take a lot of different forms. You can be inspired by anything," he says. Daniel could be happy doing many things other than music so long as they were in a creative field. He thought about this more and added that no matter the job, he would enjoy it because of his ability to find humor and beauty in almost every situation. He feels strongly his top two strengths are core to his being and happiness. Beauty is everywhere, and especially in the music he creates and mines from others.

Leadership (Justice)

Daniel says as a producer, he reads vibes and people well. Vibes can be described as the emotional aura of a person, room, or situation. He believes being good at this is fundamental to leadership. Along with having the ability to gauge people, a good leader can discern when they are needed and when they are not. Knowing whether to add something or just stay in the background is key. "You have to take your ego out of it." This is not to be confused with humility or not having personal pride. The ability to innovate, experiment, listen to your gut, and pull greatness out of collaborators comes from a pureness that doesn't involve ego.

Hope (Transcendence)

"If you don't have hope, why go on?" Daniel asks. "There's no point in NOT being hopeful. If you think everything's negative and it's all going to be worse, why are you here? Why go on?" Optimism is a source of his inspiration, "KNOWING, not just hoping, that things will get better drives me forward." Knowing involves finding out how to activate hope. He feels his best days are when he feels on the same page as the universe. That flow state inspires him to say yes to things and accept all outcomes as growth.

Judgment (Wisdom)

Judgment is staying open and considering all the evidence to make new decisions. He related this to the discovery of gravity. "If people didn't look at evidence and question, none of these revelations would have happened. How could you deduce that gravity exists, and we live in a heliocentric solar system, without taking appropriate evidence into account." Judgment is used all of the time in studio situations, and he believes it holds hands with social intelligence – knowing when to push people and when to pull back is key to infusing trust and greatness in the work.

Marni Wandner Ashby

CERTIFIED HOLISTIC HEALTH COACH • WELLNESS CONSULTANT
**Humor • Curiosity • Appreciation of Beauty & Excellence
• Gratitude • Perspective • Love**

Best-Self Story Marni's best self story was the first time she met with a health coaching client. Most people who come to her either don't have enough energy or feel they're stuck with unsuccessful weight loss journeys. They are at the point where they need someone to assess and help them. Though the exhilaration of the first time she helped a client was remarkable, she feels at her best every time she works with her clients. She gets a surge of excitement when she gets to examine their health history and issues they're dealing with and subsequently connect the dots. She wonders if "A" connects to "B" somehow and sees it all like a large puzzle to assemble. Marni is exhilarated when she has those back-and-forth discovery conversations with the client and gets to further assemble the pieces and deepen the conversation. She feels gratified when the meaningful small changes she suggests lead to some movement along their path that is visible to them – it's so fun for her!

Strength Spotting There were some clear examples of Marni's top strengths showing up in her story. Marni exhibited strong curiosity in the way she put smaller pieces together to make big-picture connections. Her perspective helps her curiosity in this fact-finding mission as she references her own experiences and past clients' histories and results. She can try that on as she looks at new cases. Also, social intelligence allows her to make connections with a multitude of different people.

Signature Strengths

Humor (Transcendence)
Marni enjoys the feeling of laughing and she's drawn to comedies over dramas. She says she can be quite intense so having a lighter side is important. Some of her humor tends to be a little dark and seeing the light in the dark is important to Marni. She was raised this way – to see the light in everything. She uses humor as a coping mechanism as she tends to make jokes in her darkest times. This core strength is one of her strongest tools that gets her joyously through good times and helps her through life's hardships. When asked if Marni uses her highly ranked humor strength as a tool to connect with clients, she replied, "Definitely – "I don't think a few minutes go by without me wanting to inject some sort of humor or levity when it's appropriate." She noted that beyond work, people in her social circle enjoy and respond to humor; "it creates a vibe and a pocket of commonality."

Curiosity (Wisdom)
Marni is a self-proclaimed digger who loves to get to the bottom of all things. She is not satisfied with the surface level of anything and always wants to know a little extra. This gives her context to figure out the world. She's always been that way and remembers asking her mother "why" all the time. The pure explanation is not enough though, she yearns to then know how it all connects. This is a great example of her curiosity holding hands with her perspective.

She also felt her curiosity drove her to recently change positions out of the music tech space. Following her curiosity led her to where she is now. She wanted the opportunity to be MORE curious and the detective work of holistic health coaching provides her with that. When asked what it would feel like to not be able to express curiosity in her life or work, she responded, "That would be like a prison. I wouldn't know what to do with myself." Our signature strengths are truly our core, and most people respond in this way when asked what life would be like without them.

Appreciation of Beauty and Excellence (Transcendence)

As a perfectionist in her own life, Marni says there's good and bad with that. "You can look at something and think 'yes! That is it!' and spot perfection. However, that can be not such a great thing when you come to expect that all of the time." She also talks about being an avid iPhone photographer. She can spot and capture the beauty in random places. Upon first moving to NY, she was awe-inspired by the composition of a pile of trash! That's someone who really can appreciate the beauty in all things.

Gratitude (Transcendence)

Marni has a regular gratitude practice that is a big part of her routine. She starts every morning this way, either writing or focusing her thoughts on small things she is grateful for. They could be simple things or even the same thing every day. She says she is not naturally grounded in gratitude, so having this practice really helps her feel cultivate this strength and connects her to the present moment. Gratitude also helps give her perspective when she feels like she could be otherwise flying in the ether.

Perspective (Wisdom)

We have seen how perspective makes cameos throughout Marni's other strengths. For her, this strength is driven by her curiosity. Perspective helps her make sense of the world which she then in turn uses to help others. "It's so much easier to coach someone else than yourself. This is why therapists have therapists and writers have editors." She enjoys taking it all in, processing it all, and then applying various perspectives to solve problems. She says good coaching is letting the client take an active role in figuring things out from the information she presents. In this way, she helps people shift their own perspective by asking leading questions like, "What if it looked like this instead of that?"

Love (Humanity)

Love really resonates with Marni. Connection is one of her biggest core values. Though this is her sixth strength, it was just too important to leave out. "Just spending hours talking with a friend at dinner, sharing and experiencing what they're feeling and reciprocating – that's just magic to me. That's love." She truly loves her friends, husband, family, and dog. The love is overwhelming sometimes. She remembers having those conversations with her grandmother growing up: "I love you so much I don't know what to do with myself!" Love is the bow that ties together all of Marni's strengths. You can see how they all work in tandem in what she does for a living.

THE HUMANITY PROFESSIONS
Love • Kindness • Social Intelligence

People who rank highly in humanity tend to see and nurture the whole. They are the session musicians who tastefully contribute their playing and understand the balance. The music teacher who takes extra time to really learn their classes' musical interests and creates lessons that do not exclude anyone. Or the social media influencer who uses their music as advocacy and understands their audience. They are savvy to people and truly enjoy a connection to family, friends, and even strangers. Here are some professions that engage the HUMANITY strengths (Table 3.12).

Table 3.12 Humanity Professions

Studio	Example Professions
A: Performance	Session Musician
B: Creation	Lyricist Music Marketer
C: Education & Research	Music Teacher (Pre-K-12) Music Professor Music Librarian
D: Industry	Music Photographer Social Media Influencer Platform/Digital Specialist

Humanitarian Professionals

Gabrielle Cornish

RESEARCHER • COLLEGE PROFESSOR

Humor • Social Intelligence • Kindness • Love of Learning • Bravery

Best-Self Story As a researcher studying Russian and Soviet history through the lens of music, Gabrielle's work has taken her to remote parts of the globe. Her pursuit of knowledge has led her down many rabbit holes (and one actual manhole) and has filled her life with adventure. Gabrielle talked about a best-self experience during a research residency in Russia about ten years ago. She joked that the city she was in wasn't too far away by Russian standards, "just a 12-hour train ride from Moscow." The *Institute for Color Music* consisted of three older Russian women who curated an archive which, she says, "was more akin to boxes of stuff."

After her daily, peaceful, hour-long walk to the archive, days consisted of digging through boxes and bags and unearthing findings amongst other interesting "weird things." She quickly bonded with these women, who would invite her to lunch and tea and generally took her under their wings. To this day, they keep in touch with her via email and wish her well every Christmas. Gabrielle had an incredible time with these ladies and communicated with them in Russian, sharing her funny research trip stories. This included the time she fell into an open manhole after dark in Siberia (which is apparently not uncommon) and another instance when she had to outrun a wild dog. After vaulting a fence to safety, she wrote her dissertation chair jokingly insisting they give her a passing grade as she just "outran a wild dog for this project." The whole experience was exhilarating. It was exciting to have remote adventures in places most people never even get to see.

Strength Spotting Clearly, Gabrielle's humor pervaded her best-self story. Not only did she use it to bond with the women at the archive, but she also used it in situations that might be someone else's worst nightmare – falling into holes in the middle of the night and being chased by wild animals. She said that even in telling this story to her worried mother, she reassured her, saying, "no, this is hilarious." This choice to see the humor in even scary situations is one of Gabrielle's signature strengths. She said that even though she is fluent in Russian, she relies a lot on humor when making connections as a non-native speaker. She is funny in both languages. Her high social intelligence was evident in the way she bonded with the older women of the archive by using her humor, kindness, and love of learning. All her strengths flowed in her best-self story. You can see the obvious bravery in the solo travel aspect of Gabrielle's work. It takes bravery to "go for it" and follow your dreams into an unknown adventure and that is just what she did!

Signature Strengths

Humor (Transcendence)
Humor, Gabrielle's top strength, was highlighted in her best-self story. She uses this tool in the classroom and her students note this as one of her most prominent traits. She uses humor to include and disarm people, especially in subjects they may not immediate connection to. Without the ability to use humor, Gabrielle said her entire trip to Russia would have been totally different. She wouldn't have been able to bond as deeply and wouldn't be able to bridge the cultural gap with as much finesse as she did. Humor has this wonderful transcendent power!

Social Intelligence (Humanity)
Social intelligence is crucial for a researcher and teacher. She believes a historian's primary job is to practice empathy and to put themself in someone else's shoes. In her own research, she has tried to embody what it would feel like to walk in the shoes of someone in 1950s Soviet Russia. Her creativity (a wisdom strength – another of her high strengths) holds hands with her social intelligence in the way she actively presents a subject and engages in research.

On the first day of her classes, she tells students that what matters to her is that they leave this class more curious than the day they entered. She believes in meeting students where they are, wherever that is, and helping them get to their personal next level. This approach of hers holds hands with curiosity (another of her high strengths) as she values the students' personal growth paths more than a standardized outcome. "Five years from now they may not remember the date that Beethoven's Ninth Symphony was composed. They can look that up quickly on their phones. What matters more is their desire to know more about the world and their place in it." Gabrielle is a great example of someone who helps others grow and cultivate their curiosity.

Kindness (Humanity)
Gabrielle believes kindness is a sister strength to social intelligence and they could not exist without the other. They are both in the humanity virtue strand as well. She points out that if two people approach an issue from a place of caring, they can have two wildly different points of view, but through kindness, they can always find a way to come together with mutual respect. She exhibits kindness as a teacher, valuing each student as an individual.

Love of Learning (Wisdom)
Clearly, love of learning was going to be at the top of Gabrielle's strengths. She became an academic so she could spend her life as a student. Her ongoing research takes her into the learning realm daily and she pays this knowledge forward as a professor. She treats learning as a sacred act and feels totally gratified when her students show their own curiosity and especially if they decide to become researchers themselves. She has encouraged and mentored many students to do just that.

Bravery (Courage)
Gabrielle is brave by many standards. Her research journeys are adventures that take her to places she has never been and put her in scenarios she has never encountered. Interestingly though, Gabrielle talks about her bravery in the context of being an ethical warrior for justice, which she says was a huge part of her development. Through childhood, she was interested in ethical issues and was keenly aware of injustices. She says bravery includes knowing your own worth and standing up for your beliefs. You can hear honesty and perseverance in this, also part of the courage virtue strand. She uses her own voice and position to help those who are being exploited or treated unfairly. This includes advocating for students and even for herself.

Paige Powell
PHOTOGRAPHER • CREATIVE MANAGER
Appreciation of Beauty & Excellence • Social Intelligence • Love • Love of Learning • Hope

Signature Strengths

Appreciation of Beauty and Excellence (Transcendence)
In Paige's favorite novel *The Goldfinch*, Donna Tart teaches us that protecting and creating art is the truest way that one can live forever; that by

appreciating art for generations, we, too, become eternal. While Paige always had a love for business, serving creativity was always at the center of her passions. "I live to protect and produce art that connects people in a world that can be very dark for those who feel alone."

Social Intelligence (Humanity)
Being a manager or creative director is all about creating a safe space for an artist. They are giving you a precious gift – by allowing you to tell their story, through your eyes or hands. You are asking them to trust you with this sacred part of themselves. As an artist herself, Paige takes pride in her own social intelligence. Through the honed skill of connecting with others, she has sharpened her ability to read social cues at a high degree. "In a crowd of people at a bar, I can always pick out the artist. The musician. The star," Paige said. "It's an energy reading skill (to learn to read the room or read the group) that takes time and more so intuition." She is quite thankful to have this skill and recognizes the many ways it compliments her craft.

Love (Humanity)
Paige feels blessed to share her life with a partner who makes her feel very safe and creative. She attributes feeling free to create and share to spending her formative years with someone who is her rock. This frees her to confidently show love, appreciate others, and express her taste. For instance, she will boldly 'direct message' creatives she likes to see if they would like to work with her. She texts sweet thoughts to her friends when she knows they need them. And she shows her love in physical tokens like birthday gifts. Paige attributes this anchored sense of love to both being secure and having a secure partner, "I have no end to the love that I can share with others."

Love of Learning (Wisdom)
Paige realized that she grew the most when she was surrounded by people and ensconced in experiences that kept her on her toes while teaching her something new. Part of the thrust for her to leave the traditional corporate structure was the "boring predictability of the day-to-day routine." She wanted to feel like she was truly learning something new every day and the rinse-repeat feel of a 9 to 5 day did not supply those qualities for her. Now, as a fully independent creative, every day is filled with meetings, calls, and projects with people who fascinate her across all industries. Her love of learning drives Paige down endless rabbit holes! "I work harder for myself than I ever could for anybody else because what I'm really working for is my personal growth and learning daily."

Hope (Transcendence)
"Without hope," Paige muses, "none of us would be here in the creative arts." She connects hope to the life of this book itself – the process of

writing it, interviewees sharing their journeys, readers dreaming of careers in music and seasoned professionals hoping to rejuvenate their creative spirits. Of all the fields, the creative ones take serious gumption! And she knows this firsthand. Paige has parted ways with more glamorous companies and clients than she cares to admit. Each time her path takes her to a moment as such, she allows herself one day-max – to feel bad or stew in regrets or "what-ifs." Each door she closes opens a new hopeful window and the lessons come flooding in. Her mentor once told her that her life would be a beautiful mural, but first she must build the wall with each building block a lesson learned. Some bricks will fit perfectly, others will be a bit cracked, others ugly and smushed in. But if you stand back, you can see the overall wall being built into a supportive structure. Paige takes heart that she is always one brick closer to her ongoing creative mural masterpiece.

THE JUSTICE PROFESSIONS
Fairness • Leadership • Teamwork

You may have someone high in the justice strengths on your team already. These people work actively to create fairness. Entertainment attorneys exhibit this every day as they make sure their clients are treated fairly in contracts and deals. Managers must be able to work the room in many different types of settings and with various teams. Band directors lead their musicians literally and figuratively through discipline and demonstrating how to achieve joint purposes. Here are some professions that engage the JUSTICE strengths (Table 3.13).

Table 3.13 Justice Professions

Studio	*Example Professions*
A: Performance	Conductor: Band, Orchestra, Choir
B: Creation	Producer
C: Education & Research	Music Administration Academic Leadership
D: Industry	Entertainment Attorney Manager: Talent, Touring, Day-To-Day, Digital, Business, Brand/Sponsorships Talent Agent Expert Witness Music Therapist

Justice Professionals

Ryan Kairalla

ATTORNEY

Honesty • Judgment • Perspective • Love • Humor

Signature Strengths

Honesty (Courage)

In his professional and personal life, Ryan tries to be honest in his dealings and adhere to ethical principles. This honesty extends to teaching a collegiate ethics class where he frequently engages students in ethical discussions. He says, "Honesty is an important prerequisite to trusting others and getting them to trust you." To that end, he is as honest as possible with others and expects it in return. His top three strengths – honesty, judgment, and perspective infuse his thoughts and actions, especially when he believes he has hurt or offended someone. He admits his wrongdoing, after taking careful time to assess his mistakes, and then he makes amends and talks about what he can do to set things right. Honesty is a core value and strength to Ryan, and the main reason clients trust his counsel.

Judgment (Wisdom)

This is extremely important to Ryan. In his professional life, he aims to make decisions that are evenhanded and deliberate. He likes to get all the facts and make informed decisions, even if that means he may not be able to decide right away. He puts himself in the position of others to see things from varying points of view. He uses this strength both professionally and personally. "I think that my top two strengths – honesty and judgment – both stem from my training and work as an ethicist. Ethics is not as much about doing what is right as it is about developing and refining a deliberative process for finding what the 'right' thing is. I think that tends to help with judgment."

Perspective (Wisdom)

Ryan jokes he's glad the VIA survey identified him as having a strong sense of perspective, "Since that is basically what I try to do for a living every day as a lawyer, I am glad to hear that this quiz confirms that I have this trait." He would be lying if he said he didn't enjoy giving advice in all realms. "I certainly like to give advice in my legal work, but I also don't mind giving advice about personal matters as well (if people ask for it)." This is very typical of people who have high wisdom – they only give their advice if asked. "I have often heard people tell me that my advice gave them new perspective on something; it caused them to look at their situation in a different way. I get a lot of satisfaction from helping others and knowing that I have some measure of tools that can be helpful. Now, if only I could be as good at taking advice from others as I am giving it! But I am a little more stubborn than I should be in that regard." This shows that Ryan's perspective is also strongly applied to his own behaviors.

Love (Humanity)

Ryan puts a high value on his relationships and he's grateful for those in this world that care about him. For example, he is quick to acknowledge the role his spouse has played in fostering his growth as a person and as a professional. "I treasure that relationship and I go out of my way to let her know often – perhaps too often! – how much she means to me." While he does regret that sometimes work obligations get in the way of spending more time with the people he loves, he continues to work on finding more balance. Clearly, Ryan has a great deal of love and passion for both his work and the people surrounding him.

Humor (Transcendence)

Ryan believes that as a lawyer, humor is a very powerful tool. He loves to make others laugh. It helps strengthen bonds with his colleagues. It is also helpful with clients. Clients typically come to lawyers when they are going through a difficult time in their personal or professional life. Sometimes making them laugh can be just what they need to help them move forward. Humor can also help in adversarial situations. When you can make an opposing counsel laugh, it helps remind them that we are all human and that our dispute is not personal. That is helpful in achieving a good result for everyone involved. You can see how Ryan deploys his humor with both perspective and judgment. All these examples further show how humor is truly transcendent. It also fuses here with perspective and helps people overcome emotions and situations that may be difficult or overwhelming. Again, our strengths never stand alone – they hold hands and meld together to make the beautiful and unique symphony called you. Ryan also uses self-deprecation humor a lot. Sometimes, people have their guard up around lawyers. But when he cracks a joke about himself, it helps put them at ease.

Shannon De L'etoile

PROFESSOR OF MUSIC THERAPY • ASSOCIATE DEAN
Fairness • Curiosity • Perspective • Leadership • Zest

Best-Self Story Shannon spoke of two music therapy stories that she says have stayed with her for years. She had a clinical experience years ago working in a neuro-rehab unit with an older woman recovering from Guillain-Barre syndrome which left her bedridden with severe weaknesses. Her treatment plan included several types of therapies (physical, occupational, and music) to help her regain her mobility and strength. There was one day in therapy that she was finally able to stand and bear weight with some assistance and the team wanted her to work on weight shifting – shifting her weight from one foot to the other which is a component of walking and getting out of a chair. Shannon began to play a waltz on the autoharp – "Let Me Call You Sweetheart" – to assist in this task. The woman's husband gets up at that point and stands in front of her. He begins to dance with her – an activity he had not been able to share with her in years. He began to cry as did everyone in the room, so touched by the special moment occurring. The focus was getting her functional and stable on her legs, and this spontaneously enabled her to regain that connection with her husband.

Her second story centered around a research project on patients with spinal cord injuries. Many times, with this type of injury, patients may never fully recover. The goal is to maximize the potential they have and make sure they are safe and functional. Shannon began working with an 18-year-old young woman who had been an athlete, swimmer, and ballerina until the age of 12 when she developed an arteriovenous malformation in her spine. This equates to having a stroke in your spinal cord, which cuts off blood flow. By the time Shannon met this patient, she was six years into physical therapy and had lost her ability to walk. Despite this, she was still an upper-body competitive swimmer. Part of her therapy was to get out of her wheelchair and safely practice walking with braces and crutches. She had a four-point gait which was very dysfunctional, and her head was down, as she feared she was going to fall. Shannon came in and used a technique called rhythmic auditory stimulation, which is basically walking with a metronome. In the first session, they increased her walking cadence by 20% and in a miraculous moment, Shannon saw her suddenly lift her head with confidence that she would not fall. This young woman had one major goal, to walk across the stage to get her high school diploma – and she did! She also went to her prom and was able to stand and be with her friends. These two stories are ones she tells when people ask, "So, what is it like to be a music therapist?" These experiences reaffirm her life's work.

Strength Spotting The act of bringing mobility and function to a person is a beautiful act of creating more fairness for that individual. Her curiosity and perspective hold hands in both stories. Curiosity drives her to really investigate each patient's goals and issues and experiment with treatments that could make a difference on a case-by-case basis. Of course, Shannon's perspective and breadth of work experience assist in this process as she can call upon past successes with patients. Her leadership shows up in multiple ways, not just in coordinating care amongst other therapists but also in demonstrating strength-in-action as she guides her patients through their journey, motivates, and reassures them that they can do it. Zest is something that is evident in the passion and emotional investment Shannon expressed in telling these best-self stories. She is a cheerleader for her patients and her positivity is contagious.

Signature Strengths

Fairness (Justice)
This shows in Shannon's motivation to be a clinician. She believes that everybody deserves an opportunity to improve, no matter how severe a condition or disability. Helping people who can't help themselves or those whom others have given up on is her core value. This is visible in her role as associate dean. In her role, she is often the frontline for complaints and issues. She has learned

over the years it's very important to get the whole story – first, listen to the person and validate their experience, then investigate the other side. This dovetails with her perspective. Patience, Shannon mentions, is a part of fairness: "Any person you may encounter, you don't know what their experience is." In both of her lines of work, this drives great empathy. "You don't know what someone is going through." Shannon is reminded of this even when she simply parks on the medical campus. "They may have received the worst news of their life." It's important for her to put everything in perspective and have empathy and patience for others.

Curiosity (Wisdom)
Shannon felt this could have been her highest strength. She has always been a curious person. Even as a child, she wasn't always outwardly asking questions, but she remembers thinking those questions all the time – why things work the way they do. This continued all the way through her education until she made the huge leap to pursue her doctorate. A mentor of hers told Shannon, "You're a very curious person; you have an intellectual curiosity and that's exactly what you need in your doctorate." The more she heard that reflected back to her, the more it strengthened her inside conviction that she wanted to feed and grow her curiosity. This is an effect of affirming someone's strengths to them – it's a powerful act of recognition and elevation. Those moments of truth tend to stay in our personal stories.

Perspective (Wisdom)
Any situation can be seen from multiple perspectives. Shannon's training as a therapist uses perspective constantly. She must put herself in the shoes of her patient and see it through their eyes, even if it's irrational or you don't agree. She says this allows her to be constantly surprised by people. Shannon's open mindset is showing up in her perspective in the way she takes each situation from a very neutral place. She does this to maximize creative solution possibilities and minimize bias.

Leadership (Justice)
Shannon is in a huge leadership role right now as associate dean, but even prior to this, there were points in her life where she emerged as a leader – and she says she doesn't know why! Many times, it was the suggestion of others that brought her into leadership. She reflects on a childhood memory of watching the Jerry Lewis muscular dystrophy telethons. They would encourage watchers to put on events locally. With her curiosity and fairness firing on all cylinders, 10-year-old Shannon decided to put on an annual summer carnival complete with games, puppet shows, magic, and refreshments. Although her parents helped, she was at the helm with a clipboard and schedule, directing the other kids in their various roles.

Many times, women who tend to be more organized and detail oriented are brought in to lead in various roles. However, a defining moment was when she came to her current university as program director of music therapy. She says this gave her tons of great leadership experience, which is not just leading the way but showing the way and developing tools to help others. Shannon's style of leadership is very humanistic, which may be why she was strength-blind to this high strength in leadership. Leadership can get a very bad rap and can be abused and done from a place of ego. Shannon's ideals of fairness and perspective make her an amazingly balanced, fair, and empathetic leader. She has spent a lot of time developing her own ideas of leadership by using her perspective and examining all types of models. "When people envision a leader, they see the person out front that they're all following or the person at the top who is telling everyone what to do and directing you. I think another way to lead is as the person at the back who is saying 'let's go, let's do this together, let's move this forward.'" She sees herself as the leader at the back, corralling everyone to move things forward. There are many roles in the arts that take this form – from stage director to choreographer – these people do a lot of their work from the back but are integral to the success of the project.

Zest (Courage)

Shannon loves that zest came out this high in her strength profile! She loved some of the questions in the survey that were asking if she got out of bed excited in the morning. She thought, "You know what, I am!" A lot of what she does in her role is streamlining systems for students to find potential pitfalls and make them more intuitive so students can be more independent and self-reliant. Constantly anticipating problems can be exhausting and wear you down. Keeping the perspective that there's a creative process in solving problems helps her amp up her zest to tackle these constant tasks. She loves that she gets to be part of making something better that will also make other people's lives better.

STRENGTHS FROM OTHER PERSPECTIVES
STRENGTH USAGE IN OTHER CREATIVE AND THERAPEUTIC FIELDS

Dawn Murnak-Karageorgos
INTERIOR DESIGNER • ENTREPRENEUR
Love • Gratitude • Appreciation of Beauty and Excellence • Social Intelligence • Zest

Dawn is an interior designer who prioritizes deeply getting in touch with her client's needs and the way they want to feel in the space. She uses her strengths, especially social intelligence, to tap into her client's strengths and desires, working side by side to create individualized, inspiring spaces for hospitality, commercial, mixed-use multi-residential developments, and custom residential design. We asked her to take the VIA survey and then reflect on her findings. Below is her written response after taking the assessment:

There is a famous quote, by the renowned and groundbreaking artist, Vincent van Gogh, "What is done in love is done well." To me, this quote embodies the notion that, when one embarks on a "labor of love," so to speak, the outcome will be seamlessly beautiful. The beauty, in my humble opinion, and possibly in the opinion of Mr. van Gogh, comes from in-hibiting the tendency to over-think, and allowing the soul to connect with the truest intention of anyone's work – which is to create something valuable, through virtue and love, for the person who will receive it.

I'm going on and on about love, not because as I sit here it is almost St. Valentine's Day, but because upon taking the VIA Character Strength Profile survey, my number one quality was Love, followed by Gratitude, Appreciation of Beauty and Excellence, Social Intelligence, and Zest. It was interesting to see that these were my top five character strengths, because they seemed to align quite well with the tools I've needed to use as a professional in a creative field. When I was asked to examine how those core character strengths impacted my work, I was quite excited to see! I was asked to also explore how these character strengths "showed up" in my life, both personally and pro-fessionally. What I found is that my signature strengths show up in pretty much all that I do.

I am the founder and principal of an interior design firm, specializing in hospitality, commercial, and residential design work. When I left my prior career in the financial industry, and took the leap into starting my own busi-ness, like most entrepreneurs would say, I had no idea what I was in for. The many years that have since past have taught me not only a great deal about my craft, the business, and my clients, but about myself. Being a business owner has taught me more about my own self than anything I've ever experienced. Followed closely by motherhood. You learn, through triumphs and failures (mostly from failures) WHO you are. The good qualities, the great traits, and the not so good parts of your persona, at least in that moment. You are utterly exposed to yourself. What you do with that personal feedback determines how far you will grow, as a professional, and ultimately as a person.

It was very enlightening, this exercise of reflecting upon my five top character strengths from the survey. Here's a breakdown of what they mean for me in the practical sense, and how they "show up" for me in my life.

Love (Humanity)

Love is both a noun and a verb, and there is a big distinction between the two. In the application of character strengths, we are speaking about the verb here. To love the process, not just the outcome. That's a goal that I always try to attain while working on a project. And just like the wonderful van Gogh quote, if you love the process, more than most likely the outcome will probably be "kinda great."

Love shows up in my professional life a lot, and not the way that sounds. But when I meet with a client for the first time, my first thought is of love. Nothing romantic, but a humanistic, compassionate love. I feel such a connection with people, and the love of people, and desire to help them, that the I want them to feel that when they meet me. I want to help as many people as possible feel the sense of happiness and beauty through my work, and that comes from a genuine sense of connection and love.

I also love the creation process. For me there is magic in creating space and environment. The highest compliment I've ever received is that someone can feel the love and effort that went into a space that I designed.

Clearly love, in general, is of utmost importance to me in my life. I am blessed to have found the love of my life. My husband and I truly love each other equally, and he makes me feel loved every single day. I've also recently experienced the love and joy of being a parent. Having welcomed a son fourteen months ago, I can say, what everyone told me was true. My son has taught me so much in his little, wonderful life at only fourteen months. He loves unconditionally. You can put him up for his nap and he's crying and upset, and then when you pick him up after he wakes, the smile on his face is just priceless. Why? Because he loves me! No matter what, he doesn't care that I made him take his nap, or made him eat vegetables, or didn't let him touch a sharp corner. Even if he gets upset with me in the moment, he still loves me just the same. And the feeling is mutual. Seeing his little face filled with delight every day, just learning the world he lives in, laughing and playing, is the definition of love and joy.

Gratitude (Transcendence)

Gratitude is something that permeates every moment of my life, every day of my life. I was not too surprised that this character strength was second on my list. I credit my mother for influencing me in this area. Growing up, she would say to me, pretty much every day, to be grateful for what we had, and to have empathy for others. As I grew older, and learned, through my own life experience, how the simple act of being grateful could transform my life, the lesson from my mother only grew stronger and more a part of my personality.

I feel a sense of gratitude with every breath I take, and every day that I'm here. I know that might sound borderline (or over the border) corny, but it is absolutely true. Feeling gratitude for the life you have, and by extension, the things you have, the people you have in your life, and the experiences that you are presented with – immediately turns problems into opportunities, feelings of lack into feelings of abundance, and troublesome relationships into life altering lessons. Gratitude is magic, and I feel like it is the key reason for why I feel happy. There's a quote by the world-renowned self-development author Wayne Dyer, "When you change the way you look at things, the things you look at change." This simple quote is the key insight into the magic of practicing gratitude.

When I have bad days where I feel stressed and overwhelmed, it is usually because I have a situation at work that is depleting my reserve of time, energy, and resources. I've had plenty of moments where I start to feel down about it, upset, frustrated. That can last a long time if you let it, and when you own and operate a business based on client interactions, those negative feelings and thoughts have the power to become contagious. Those bad boys can spread. They can start permeating your communication with clients and colleagues. Who hasn't received an email from someone where you knew exactly what type of day they were having? Those thoughts can soon turn toxic and have the power to start pushing you down a rabbit hole that can easily turn into a spiral, the downward kind. The antidote, in my humble opinion? Gratitude. I turn my mind to this thought – I chose this life, I chose this path. There is nothing here that I didn't choose, and therefore didn't anticipate could happen. I'm grateful for the opportunity to choose this path. Clients, and therefore client "issues" are a blessing, because they mean that I've been blessed to pursue this path, that I chose.

See? Pretty simple. Just the shift in perspective, and the sprinkling of a little love-gratitude-magic dust, and the downward spiral just turned into an upward smile.

Appreciation of Beauty and Excellence (Transcendence)

This character strength seemed pretty straightforward to me, as one of the tenets of the interior design field is to have a love and appreciation for beauty and excellent craftsmanship. I can recall, decades before I ever had the hint that I would end up in a creative field, when I was a little girl – probably eight or nine, I would say the ONLY thing I wanted for my birthday – and keep in mind the age – was a Pergo floor in my bedroom. Gone with the sea foam green "sustainably made fibers from recycled cola bottles" carpet that my parents had installed (and in hindsight was probably the real cause of my horrific childhood allergies), and in was a blonde "wood," light-grained floor that made me feel like I was in a new world, let alone a new room. In addition to that floor, I begged that we paint my room a very pale shade of lavender. I got my ideas from perusing an IKEA catalog, and seeing the simple, Swedish aesthetic, and wanting nothing but.

Even at a young age, the appreciation of beauty was strong. As I grew older, I was influenced greatly by the outside world asking, "What are you really going to do for a living?" and didn't really acknowledge the fact that a job or professional could actually revolve around something other than stocks or trades. After high school, I attended college intending on a business degree. In fact, I did receive one (I have a bachelor's of science in economics) but that "appreciation of beauty" side of my personality could not stay hidden. As I toiled with Game Theory and macroeconomics classes, I secretly started pursuing classes in art history. One class can't hurt, right? I was always fascinated with ancient culture and art ever since watching Indiana Jones dash around a cave amongst ancient Egyptian totems and relics, back in the 1980s. Wasn't the point of college to bask in something that you were interested in

(No offense to John Maynard Keynes)? So, I took another art history class, and another. And soon enough I had enough to complete an art history minor. Not only the beauty of the ancient world, but the sheer excellence of their artistic and architectural works did something to me. It inspired me. It made me feel small, in the most amazing way. It fascinated me. I should have known that I would eventually leave my career on Wall Street to start an interior design firm. And I'm fully aware that if I had been born in the year 2500 BC, neither of those options would have been possible for me.

If I stop right now and realize that I have a career in a field that revolves around beauty and excellence, it is exhilarating to me. Like, "catch my breath, how on earth did I get this lucky" exhilarating. Creating beautiful spaces that evoke emotion is one of the most amazing opportunities that I can imagine. And the space isn't even for me!

Social Intelligence (Humanity)

Client-based businesses revolve around social interactions. Social intelligence is a highly valuable asset if you are engaging in daily communications with vendors, clients and colleagues. And these days, it is hard to avoid the daily communications, whether in the form of emails, texts, or in-person engagements and meetings. Social intelligence helps grease the wheels of those communications.

It also helps in micro ways: hearing a little inflection in a client's voice or empathically understanding what they might not be able to iterate or explain in words. Or to know what they want without them even knowing and delivering on it. All of these take social intelligence. Sometimes the noise of the everyday can block that ability. When that happens, I try to just be in that very moment with the client. I listen carefully. I hear the subtext, not only the words. I tune into their body language and engage myself in the moment. These exercises have helped me hone that skill and trait.

Zest (Courage)

Last but not least, zest. I'm assuming this character strength is zest-for-life adjacent, so I will proceed with that understanding. I guess I can see how zest would come up on the top of my strength list, and frankly, it makes me happy that the survey could see through the layers of stress, responsibilities, and to-do lists to see through to a trait that I was once only characterized with. When I was younger, from school days (where I was head cheerleader) to always being called the cheery, smiley one in every waitress position as a young adult, zest was my claim to fame. And it wasn't fake – it was legitimate. I just had a sunny disposition. I still had a lot of zest in my career in the financial industry. I enjoyed my work, came in early, worked late, and still was zestful all the way home.

Thinking back to when I started my own business which, as I mentioned, showed me so many facets of myself – the good, the bad, and everything in between. All that growth and knowledge is fine. However, the other part of starting a business wears you out, or at least, me out. I can't speak for everyone, but most business owners I am friendly with have felt the strain of burnout, and some (myself included) have never taken the time to un-burn ourselves. So as one might expect that poor little thing called "zest" can be a casualty of this burn out. I hope that has not fully been the case for me, but I have to admit, I don't believe my zest meter has been as high as it was back in those cheerleading days. But that's why taking this assessment and seeing this result was so illuminating, and inspiring. That my zest is not only in there, but also actually prevalent. So prevalent that it holds a top position!

What a wonderful thing to realize, that no matter how much you think you've abandoned some of your most cherished traits, they are still very much in you. And isn't that the true gift of self-reflection, and the actual reason for taking the evaluation? To understand that whatever top character strengths the survey reveals, you already know what they are deep inside. And though life can cloud and complicate your self-opinion, something as simple as answering a few questions can help you hold a mirror up to yourself.

The outcome, those little words in black and white, can gently redirect you back to your true path, the path of who you are and what you are meant to do, whether you've strayed just a little from that path, or if you were lost in the woods. It's that little reminder that the character strengths you hold are there always, and the world is just waiting to receive them.

Phoebe Atkinson
LSCW • PSYCHODRAMA TRAINER • COACH • PROFESSOR
Gratitude • Love • Curiosity • Honesty • Love of Learning • Zest
• Appreciation of Beauty and Excellence • Leadership

Phoebe was first introduced to the VIA strengths survey when she enrolled in the certificate in positive psychology program at Kripalu in 2012. She took a deep dive into character strengths right from the beginning as she saw their connection to a virtuous and meaningful life. She took VIA courses online, did strengths exercises, and did several 30-day challenges using them.

For Phoebe, learning about character strengths felt very affirming as they named her most natural attributes. She joyfully owned that there is 'no one else like her'. It was an uplifting, touchtone to her essence that provided her with an important entryway. Phoebe also likes the VIA aware-explore-and-apply teaching model as it creates an organic pathway to flow states.

Phoebe looks at strengths in terms of the top ones, the phasic (situational) ones, and the lower ones. She believes that to cultivate strengths, one must be intentional, make a plan, and practice until it becomes a habit. She employs "towing," a process where you use your top strengths to move and grow other less developed ones. Phoebe builds her courage strengths using her zest, authenticity, and especially her leadership. Lately, she has been elevating her spirituality by using her love of learning and researching "awe" for an upcoming presentation.

Phoebe views strengths as many colors of the rainbow, which we all have inside of us. The rank order aspect is not as significant to her in her teachings. She uses the Segura Color Mat, which places all 24 strengths on a horizontal plane. Phoebe is in the creative stage of working with character strengths. She thinks about her clients before she meets with them and tunes into their top strengths. This neutral, nonpathological way of working allows her to language things through a strengths lens, while also teaching the strengths all along the way. If she needs to, she can dial a strength up or down to find the "sweet spot" or address over and underuse.

Phoebe's deep and natural connection to gratitude and love is front and center of who she is – you feel it when she listens to you intently, teaches with passion, and when you see her in action facilitating a group. She lives and breathes gratitude and love in all her relationships and interactions. "I love gratitude. I wake up every morning with such an appreciation of life and where I came from. I am the daughter of a marine who was educated, dedicated, and served our country. My mother was a college graduate of Bryn Mawr." If you would shadow Phoebe's zestful life, you would witness her many roles – sister, parent, mentor, coach, psychodrama trainer, therapist, writer, professor, and dear friend to many. Phoebe loves life and lives it from a lens of appreciation while giving service generously and graciously.

Her communication skills are finely honed, and she is constantly connecting and extending people's social and professional networks by introducing everyone to each other. Phoebe often teaches about the "reciprocity ring" as she is keenly aware of and grateful for how people have shown up in her life when she needed them, to guide and help along the way. In relationships, she shows mutuality by both giving and receiving.

Phoebe also has many intellectual capacities and gifts. She is a go-to for her current knowledge about positive psychology – the important writers, and the recent literature and research. Not only does she remember many details, but she also has a unique ability to systematically interweave and integrate many different fields of knowledge. Love of learning and curiosity are part of Phoebe's DNA and motivate her like bright headlights leading her down many paths. Her curiosity is like a beautiful golden thread enticing her to go deeper, ask more questions, and learn more. She is always noticing what is new and pushing her personal "refresh" button for new experiences.

Phoebe is loyal, fair, and trustworthy – this is her honesty. In the past, sometimes honesty was a stretch zone for her – more in terms of giving herself the permission to speak up. This has shifted over time through experiences and personal work. In the past, focusing so much on others and being in service could create an imbalance that didn't allow her honesty to show up as clearly as it does now.

Phoebe calls zest the "miracle grow" of character strengths. She is passionate about learning and teaching as a prelude to sharing this knowledge with others.

Phoebe was "strength spotting" before most of us even knew about character strengths! Strength-spotting exercises are used to help people identify their strengths and the ways they use them. Imagine being seen when you are at your best, using the strengths that light you up and engage you, and having someone mirror those strengths back to you by naming them and giving them agency? It's like a refresh button at the soul level! Phoebe has integrated this so deeply that she naturally speaks the language of strengths most of the time. Character strengths interventions are the most highly applied skill of all positive psychology interventions and Phoebe is a pro in this arena.

For Phoebe, the appreciation of beauty and excellence ties into several of her life paths and passions. She is always looking for the good and striving to make things better.

Phoebe sang opera as a young woman and continues to have a deep relationship with music and the arts – including theater, movies, and literature. Her excellence shows up daily in everything she touches – her facilitating live or online, her teaching, coaching, and clinical work. Phoebe often reminds people that character is plural, and strengths work as a team. She joyfully prepares everything she presents. It's the resulting teamwork of her curiosity, love of learning, zest, and excellence.

In this way, Phoebe works with the strengths as a "combo platter." The combo platter includes all our strengths, used in varying degrees, and sometimes over and underused ones. Strengths show up in the many roles we play, and it's all context dependent. She listens for people's signature strengths to figure out how to best motivate them.

Love is often one of the signature strengths of good leaders, as well as love of learning, teamwork, and leadership. Phoebe has three of these strengths. Leadership is a strength that Phoebe "wears" well. Combined with her gratitude and love, she beautifully represents "the servant leader." Her mindset is "serve-first," prioritizing the needs of the organization, group members, or students in an honoring and responsible manner. All the leadership roles she desired required her to do new things and make brave and bold moves.

Phoebe is known for being a terrific collaborative partner. During the pandemic of COVID-19, she curated over 300 lunch and learn workshops for Whole Being Institute in collaboration with JCC in Manhattan. She co-led and provided support and technical input to many of these. What a combination of love, leadership, and gratitude during a time when we all needed to stay in touch and keep growing!

Growing the strength of hope is important to Phoebe and for her clients as it embraces both optimism ad future-mindedness. In her work coaching and as a therapist, setting goals is an important aspect. Hope is a key factor in setting goals as it relates to flourishing, zest, and gratitude. One of her mentors, Don Tomasulo, wrote a recent book, *Learned Hopefulness: The Power of Positivity to Overcome Depression*, which Phoebe uses as a great resource for building hope.

Phoebe loves the "Zen of the VIA," as there is always something new to see and learn each time you work with your strengths. She believes to live a virtuous life you must intentionally apply your strengths. They are always there waiting for you to ignite them and use them to engage your best qualities. They are like a magic pot of gold in which you can always dig deeper.

"My strengths are my superpowers. They explain what makes me tick, how and whom to partner with, and how I can be at my best." Phoebe sees herself as an ambassador of knowledge – she receives it only to pass it on. She quotes Luke (12:48) "To whom much is given, much will be required." This wisdom holds us responsible for what we have.

Phoebe is grateful for her given talents and strengths and generously shares them to benefit others. This is one of the stated missions of positive psychology – to take all the learning and research to "Main Street" and the people. Phoebe does this and embodies positive contagion as she shares her talents generously with everyone around her. To be in the sunlight of her gentle and loving essence is a real gift.

Megha-Nancy Buttenheim
YOGA TEACHER AND TRAINER • POSITIVE PSYCHOLOGIST
Appreciation of Beauty and Excellence • Gratitude • Humor
• Teamwork • Zest • Love • Hope • Spirituality

Megha was introduced to the VIA character strengths in 2011 when she was assisting Whole Being Institute in designing the first positive psychology certification program, CIPP (Certification in Positive Psychology). She has since been delighting many communities with her creative use of strengths.

When she first took the VIA survey, she found the questions fascinating and she was curious about the results. Megha was blown away by her initial strengths survey as her top strengths were such a good fit. That moment ultimately proved to be a life-changer that helped her see herself in a hugely new way. It boosted her confidence and healed a lot of her insecurities. Since that moment, Megha has taken the VIA several times and her top three strengths have remained the same. Four of her top eight strengths are in the transcendent virtue category.

The strengths are a regular part of Megha's everyday life. She sees the VIA strenths as integral to positive psychology and the sciences of happiness and resilience. "The VIA is one of the best tools when I am going down. Even on glum days, the VIA strengths pop up reminding me who I am."

Megha soon after got certified in positive psychology and began thinking about how to incorporate the strengths into the body with music and movement. It always felt natural to her to do so, and she intended to teach it in that manner – she wanted people to feel the strengths in a visceral way. Thus began her deep dive into each of the 24 strengths. She took time to first internalize them. Next, she found songs, movements, and poetry to help people kinesthetically embody their strengths in their bodies. She listened to a lot of music to find songs that resonated with the strengths and then asked herself "What is the best movement for this strength?"

Since 2012, Megha has been weaving strengths into her work and training sessions. At times it was frustrating because no one was systematic with the bodywork until Megha came along – even though the concept of whole-being included the physical. She knew she was creating something good and important, so she persisted.

In her many years of teaching, Megha based much of her work and her Let Your Yoga Dance series on the chakras. Chakras are invisible centers of subtle energy or prana (life force) aligned alongside the spine. This system is ancient, with its origins in India as far back as 1500 BC (*What are chakras? concept, origins, and effect on health*, 2022).

Megha's Let Your Yoga Dance series is a creative fusion combining dance, yoga, the breath, quotes, relaxation, savoring, and curated music in a joyful and fun way. It is user-friendly and intended for people of all ages and abilities. Along with addressing the body, mind, and spirit, Megha also incorporates positive psychology concepts about love and gratitude. Prompted by Dr. Tal Ben-Shahar, she turned this into a book *Expanding Joy: Let Your Yoga Dance* (Buttenheim et al., 2016).

In 2018, Megha created and launched "Moving with Your Strengths," a series of 30 short videos that teach ways to embody our strengths beyond journaling and standard exercises using the gifts of her years training in yoga, meditation, relaxation, chi-gong, dance, and movement. Megha created this project as a companion piece to learning positive psychology and the main pillar of character strengths and she is very proud of it! The 24 VIA character strengths provided Megha with another research schema to articulate each strength. It was another lens to teach through.

Megha's teaching is unique in form and manner of delivery. She sets the tone with kindness, compassion, and humor. She embodies gratitude and encourages students to be grateful as well. Megha coined the term "blessons," her original combination of a blessing and a lesson. She doesn't believe in critical feedback when teaching. A blesson is offered in a positive way with a tone of excitement, kindness, and receptivity. For example, "Next time, let's try this and see how it goes."

Additionally, Megha offers a 21-day practice three times a year called "Moving with Your Strengths." Inspired by the many permission mantras in positive psychology, Megha's mantra is, "permission to be."

Megha reflected on her top five strengths. Her number-one strength, appreciation of beauty and excellence, gave her a new frame for her focus and attention to detail and allowed her to see herself as an artist. She had always labeled herself in a negative light as persnickety about how things turned out. What she had judged, she now saw as an asset, a strength.

Gratitude was already a practice in Megha's life as she had kept a gratitude journal every day for over forty years! She also combined gratitude with her prayer practice. "Being a grateful person is who I am."

Seeing that humor was her third strength tickled Megha and gave her a lot of confidence. She shared, "Boy, that's a healthy strength" as she laughed her easy, rich laughter.

Teamwork comes easily to Megha, "I'm a great team member – I want to lift people up!" This is evident to all who work with her.

And zest is a no-brainer for Megha as she has a lot of energy – just watch her! It has been an adjustment, post hip replacement, and she has a slower kind of zest now.

This surgery really activated her strengths. This was Megha's first-ever surgery and she was scared. She used her strength of love and leaned on her relationships right from the get-go. She sought out support and information from her tribe and really rallied her courage and bravery. After the surgery, she became grateful in a different way for new things she notices, "I'm grateful for my new mindful cardiologist Dr. Cippy. She mentioned in our visit that I can heal this through my emotions! My kind of doctor!"

She realized she never thought much about her heart and now she pays attention and sends it love. She has a new awareness of other body parts hurting, "I am taking care of myself at the highest level I ever have." Prudence (not taking undo risks) was one of Megha's lower strengths. While normally she didn't think much about this strength, being careful became very important to her.

Strengths that Megha is actively working on growing are forgiveness, honesty, courage, and leadership. She is working on letting go sooner rather than holding a grudge when someone has been hurtful or unkind. This has been challenging. In developing her leadership strengths, Megha is in a new stage of mentoring. She is focused more on the training of the trainers to carry her work forward. With newly found courage, Megha now clarifies her boundaries and can articulate when a group member is complaining and bringing the group down. This goes hand in hand with her increased honesty. She used to have a list of things she tolerated but now she speaks up more. Rather than putting up with things, she names them clearly to the person.

Megha personifies "living and moving with your strengths."

Megha-Nancy Buttenheim, M.A., is the founder of Let Your Yoga Dance: Grace in Motion® and an international presenter, Megha has been leading teacher trainings, retreats, and classes at Kripalu Center in Stockbridge, Massachusetts from 1985–2019. Since then, Megha has been a consultant and teacher for many programs – WBI, JCC Manhattan, Dr. Tal Ben-Shahar's program as well as having an independent private practice. Megha coined the term chief joy officer (CJO) to bring home the fact that we are the chief joy officers of our life and work.

Simone B
A STRENGTHS STORY
Love • Perspective • Social Intelligence • Humility • Appreciation of Beauty and Excellence • Prudence • Honesty • Humor

Simone lives a full life. She is a mother, a grandmother of two small children, a sister, a beloved friend, a choir singer, and at heart, a creative – formerly designing and dressmaking. There is a nine-year span between her VIA survey in 2014 (Social Intelligence, Honesty, Love, Humility, Appreciation of Beauty and Excellence, Teamwork, Curiosity, and Humor) and her most recent one in 2023 (see above). Dr. Nancy Kirsner introduced Simone to the VIA. When she first took this survey, she was returning to the workforce after many years, and getting her sea legs by taking on a new library assistant position in a large local hospital. This new learning curve was quite challenging and the job required her to use computers.

When Simone was first introduced to the VIA survey, she "hated it." It brought up all her bad feelings about taking tests as a younger person. At this time, she saw things as black and white and believed, "Tests were written for extroverts like Mary Poppins, not for introverts like me." She saw these tests as very binary – you were either good or bad. She believed the wording was written for the business world with criteria for evaluation that she did not fit. Tests brought out her worst feelings from childhood eliciting feelings of being less than others and feeling dumb. Simone explained that in her home country, France, they would read the results of tests out loud in front of the entire class and it was so shaming. This became her closed mindset, the old narrative of thoughts, feelings, and beliefs.

Even though it challenged her, she completed the survey out of love because Nancy asked her to. Over time, with a lot of personal work, and through an ongoing strengths practice, her mindset and way of seeing herself began to shift. The magic of her strengths rainbow was finally emerging for her to see and apply.

Simone's black-and-white thoughts shifted to shades of gray as she saw, owned, and valued her strengths. Life became more colorful, and she learned about concepts like going from "wrong to strong" and "good enough." Simone also learned about the negativity bias and used her growing strength of perspective to help offset this. She realized when she was tired and at the end of the day, her black-and-white thinking was the negativity bias in another form.

As she saw her strengths more clearly, she began owning her value to people. For example, when her trainer moved into her building with his two younger boys, both would seek Simone out for company and to talk. One of the sons wanted to learn to sew and Simone was an expert sewer and was delighted to give him lessons. This relationship blossomed and Simone felt the joy of teaching him enlivened her life. Fast forward seven years and that young man applied to a specialized high school program for design. In his application essay, he wrote about Simone's impact – not just by teaching him to create patterns and sew, but also about how she was receptive and a great listener. She was very touched by this and even commented that he will be her creative and design legacy, as none of her own children expressed an interest in this.

Social intelligence was always Simone's "go-to" strength. People with no filter sometimes triggered her, so she started using gratitude and perspective (two strengths she worked hard to cultivate) as an antidote – being thankful for her humility and ability to contain. Gratitude now feels like second nature and she experiences her entire life through its lens.

Perhaps the largest impact of Simone's strength practice is that she no longer sees the world as black and white. To help achieve this, she steeped herself in mindfulness classes and worked with a mindfulness coach. This was a tremendous boost that helped her be more present and see things through a neutral lens. She now sees her mindset as open and growth-oriented and feels empowered that she can learn new things, practice, and make changes. Character strengths were key to unlocking this door as Simone persisted, despite the difficult feelings that just about learning the strengths brought up for her.

Simone's strengths became more organic to her at the beginning of the pandemic. She knew that COVID-19 was going to provide lots of opportunities for positive growth and change for her, even early on when we were all pretty shocked and upset. She began designing different kinds of masks out of pretty fabric she had, way before they were available in the community or online. She distributed them locally to a retirement community and to other individuals.

Simone also focused on growing her teamwork strength by joining a choir and volunteering at several places working with many kinds of people. One of Simone's greatest takeaways was learning she doesn't have to be alone and do everything herself. This dispelled an old childhood myth that she had to carry the weight of the world on her shoulders. She could also validate others and they could do the same for her. She learned about reciprocity in action!

Today, when Simone is in difficult situations, she owns and knows her intelligence, and uses perspective to explore beyond the options and choices immediately available to her. Learning the actual language and definitions of the strengths was most helpful to Simone, rather than the virtue categories.

By comparing the initial and recent VIA survey, Simone has some insights into the ways she has learned to apply and analyze her strengths. In the last two years, she moved closer to her oldest daughter and saw the birth of her first two grandchildren. This new role as grandmother has elevated Simone's love to a new high. Prudence rose many ranks in her strengths (from #12 to #6) as many transitions have required a lot of decision-making involving finances, selling and buying a home, retirement, moving expenses, contracting, and rebuilding a life in a new city. And Simone now fully believes in herself and owns her wisdom in her roles as friend and grandmother. She realizes and appreciates how much she knows and must share with others.

Simone isn't always aware that she's working on her strengths. However, when she is going through something, and looking at the strengths on a page, she does realize there is always work being done. Currently, she is working on forgiveness and the power of letting go. Simone, like all of us, is a beautiful work in progress. Once you open the strength's treasure box, you will be surprised and excited about the gems inside.

Second Set Practice Room

Choose someone in your field whom you consider a success. Identify and write about the top eight strengths they are demonstrating and tie these strengths into the "how" of their success. How do these strengths contribute to their success?

_____
_____
_____
_____
_____
_____
_____

References

Buttenheim, M. N., Ben-Shahar, T., & McDonough, M. (2016). *Expanding joy: Let your yoga dance: Embodying positive psychology.* Grace in Motion Books.

Keltner, D. (2022, January 10). *How to find your best possible self.* Mindful. Retrieved January 30, 2023, from https://www.mindful.org/how-to-find-your-best-possible-self/

MediLexicon International (2022, May 24). *What are chakras? concept, origins, and effect on health.* Medical News Today. Retrieved February 26, 2023, from https://www.medicalnewstoday.com/articles/what-are-chakras-concept-origins-and-effect-on-health

Tomasulo, D. J. (2011, December 27). *The virtual gratitude visit (VGV): Psychodrama in action.* Psychology Today. Retrieved January 5, 2023, from https://www.psychologytoday.com/us/blog/the-healing-crowd/201112/the-virtual-gratitude-visit-vgv-psychodrama-in-action

Wikimedia Foundation (2022, November 27). *Values in action inventory of strengths.* Wikipedia. Retrieved January 5, 2023, from https://en.wikipedia.org/wiki/Values_in_Action_Inventory_of_Strengths

CHAPTER 4

Last Set

Apply – Strengths in Action

Last Set

1. *noun*: the final set the band plays at a show. Sometimes this is followed by an encore.

The Importance of Practice

What if you could find small ways each day to optimize what you do and feel energized while doing it? There is nothing difficult or costly about focusing on your strengths. They are your daily "nutrients" for a life of well-being. Plus, making a commitment to do so will shift your whole point of view to one of a growing and open mindset. You will feel and be different. And no one must preach to the choir about "practice making perfect." The more you utilize these strength muscles, the stronger and more natural they become to help you move through your day, career, and life.

We have gone through two of the three stages in the VIA aware-explore-apply model. In this section, we will show you specific activities to apply these strengths in your own lives, personally and professionally, and to use them as tools of empowerment.

Working with your character strengths is one of the most exciting and satisfying areas of personal work and growth. There is no one way to go about this, so let your imagination and creativity guide you in these experiences. Learn and experiment! Make up your own strength cultivators! You can think of this as learning or listening to new music.

Remember, there are no absolutes or final conclusions. It's about "having new eyes" through which to see yourself, others, and the choices that you make. Over time, as you continue a character strength practice, you will

DOI: 10.4324/9781003267607-4

deepen the layers of your self-understanding and lean on your strengths to overcome struggles and optimize your performance, and live a happier, engaged life of well-being.

The Payoff of Practice

Need some more motivation to get going? Or are you trying to motivate your squad? Here are some quick incentives to start your practice today:

- Using your strengths makes you happier. This is linked to lower levels of depression and higher levels of well-being, vitality, and good mental health (Haverson, 2018).
- Using your strengths gives you more energy and vitality leading to an active life including eating well and engaging in enjoyable leisure activities (Rothstein & Stromme, 2018).
- Using your strengths increases your confidence which is associated with some powerful "selfs": self-efficacy, self-esteem, and self-acceptance (*Build your confidence by identifying your strengths*, 2022).
- Using your strengths leads to higher levels of positivity. The strengths of kindness, social intelligence, self-regulation, and perspective are especially helpful as a buffer against the negative effects of stress and trauma (Wood et al., 2010).
- Using your strengths helps you personally grow and develop. Strength building, which involves positive self-monitoring, increases self-development when you're learning something new or something difficult. This approach also helps the learning and changes stick (*Focus on your strengths, focus on success*, n.d.).
- Using your strengths improves your confidence in a relatively short time (Dalla-Camina, 2022).
- Using your strengths increases your life satisfaction, which leads to good problem-solving and better performance. People who use their strengths are more likely to achieve their goals, and with more reported satisfaction (Clarke, 2018).
- Using your strengths results in better physical health and makes you more resistant to stress (Zhang et al., 2021).
- Using your strengths creates a life whereby you experience more meaning in your work and relationships. This leads to more job satisfaction, pleasure, engagement, and meaning. Studies have found that employees who use their strengths regularly at work each day are up to six times more engaged in what they're doing (Asplund et al., 2023).
- Using your strengths helps you perform better at work. You are more creative and adapt better to change with proactive behaviors based on attention to detail and hard work. Emphasizing personal strengths, rather than weaknesses, results in higher performance (Roberts et al., 2022).

- Using your strengths is correlated with more resilience and adventure. Leveraging your strengths allows you to travel outside your comfort zone, which in turn builds resilience (McKay & McKay, 2021).

GENERAL STRENGTH ACTIVITIES

Warm It Up!

To cue yourself into the energy of this practice, you can start by doing these prompts designed to warm up your positivity and prime your brain for savoring things that make you feel most alive!

- I am happiest when …
- I feel hopeful when …
- I get excited about …
- I take pride in …
- I am proud of …
- I am interested in …
- This is something that touches me …
- I am inspired by …
- I am grateful for …
- I am amused by …
- I am in awe of …

Daily Strength Rituals

Repetition is key to habit formation. It is especially so when attempting to shift our hard-wired negative bias to a more positively centered mindset and way of seeing and talking about ourselves. Daily strength spotting is key to absorbing the language and practice of the strengths. The more you notice, name, and actively use your character strengths in daily life, the more you are training your brain to do these things naturally. This is called neuroplasticity. Our brains can grow and change throughout our life span. Here are some daily practices to adopt:

- Start each day by affirming one of your strengths in the mirror. Celebrate it out loud!
- End each night by acknowledging which strengths you used in the day and how they helped you.
- Journal your day through the lens of your strengths. Instead of making a rote list of events that happened, be specific about how the strength showed up.
- Pre-game your week by calling upon your strengths the way some people call upon angels. They are here to power you through, and

they are in your DNA. They are always with you. They are your strong power source.

- Take the time to notice strengths in someone close to you and compliment them – let them know you see them!
- Express daily gratitude for your strengths.
- Express daily gratitude for the strengths of someone you meet in the day.
- Create a calendar of the strengths by writing a different strength on each day of the month. Decide to see each day through a different strength. Journal how that strength showed up. Get creative! You'd be amazed at what you see when looking through a new lens.
- Consider a strength you will need to handle an upcoming situation. Write a strength down on a piece of paper or on a note on your phone and look at it as often as needed.
- Wear your strengths on your sleeve! Well, sort of. Wear your signature strengths on bracelets as daily reminders that you are powerful! (See Merch Booth.)

Group Strength Activities

Groups can have the ability to bring out different parts of our personalities and help us see ourselves through the eyes of others. They can consist of strangers or our loved ones. Some group activities help to really affirm your strengths. When you are in a group with people who know your personality, it is fun to hear how they see your strengths show up in you – even strengths you didn't consider or are blind to. Here are some group activities to try:

- In pairs, tell a best-self story. Take two minutes and describe a time when you were at your best, happiest, feeling in flow (time flying by without you even noticing), or feeling like you were doing what you were meant to do. As you describe this story, your partner will name the strengths they heard. You can also add in any other strengths you noticed. It's helpful to use strength cards (see Merch Booth) for this activity to both keep track of your strengths and concretize them with an object. When you are done, reverse roles and listen to your partner's story.
- Using a strengths mat (see Merch Booth), warm up a group of people by having them all stand on their top strength. Everyone then tells how this strength shows up in their daily lives. This becomes fun, like a game of twister. It is also interesting to see how the strengths are spread into different virtue categories. It's connecting and fun when multiple people have the same top strength and try to stand on the same spot. This is a wonderful warm-up and inclusion activity because a group of strangers

automatically have something in common to discuss and it's a safe way to get to know each other.

- There are many other ways to use the strengths mat, including a strengths "walk-about." Have everyone in the group stand on each of their signature strengths and describe how those strengths show up for them one at a time in multiple rounds, going through the top five. This is a great way to get deeply and quickly acquainted. Think of all that is gained by this line of questioning rather than more surface inquiries – it skips the small talk! A variation of this basic strengths walk-about is to name your top five signature strengths in music or your arena of work.
- Start a group chat for strength support. Each day, declare a strength you are going to bring into the day. Encourage others by reminding them of their strengths. This could pair with any other cause or challenge – fitness, academic, or fundraising groups. Your strengths can fuel your efforts in all areas.
- Create a 30-day challenge around character strengths. Thirty days is about the time it takes to establish the beginning of a habit pattern.

MUSIC STRENGTH ACTIVITIES

Music making and working in the field of music at all levels is a very different endeavor than other professions and activities. Music itself activates just about every part of the brain (Budson, 2020). Therefore, it is such an amazing therapeutic tool, formally in music therapy and informally in your relaxation, energizing, or break-up playlist. One can easily understand the lure of the muse (the transcendence strengths align with this). Although music has many entry points, those who really throw themselves into studying it know the prodigious amount of work and discipline it takes to get great at it (temperance strengths like self-regulation, prudence, courage strengths like perseverance and zest, and wisdom strengths like creativity and love of learning). The "virtuosi" out there can list the many times they did not engage in other social activities, in favor of logging hours in a practice room or preparing for a performance. And the stamina it takes to get to and stay on top of the charts is equally formidable.

Musicians are often told to cultivate very contradictory behaviors and perspectives – be open and vulnerable in your expression and performance but at the same time develop a thick skin to criticism and by the way, odds are you won't make it. This is not talked about enough in the field of music. These diametrically opposite concepts and energies are supposed to somehow coalesce. Those who work with musicians at all levels need to be aware of these struggles that are not often discussed. Learning to hold the tension of the opposites and deal with these nonlinear concepts is important. It is one of the main motivations to write this book.

People drawn to music tend to be on the more sensitive side of things – they are tuned into the subtleties of life and beauty. And contrary to popular belief, not all performers are extroverts. Many very famous performers are introverted and find the spotlight of the stage almost painful. People who go into the field of music performance do so against the odds of "success." There are careers that, upon completing education, dovetail into the professional field with expected beginning salaries and a conventional growth path. Music performance is not one of those. This path is paved with hard work, luck, the right contacts, the right opportunities, and the flexibility to do many things – even those you are uncomfortable doing and are learning along the way. It takes tremendous perseverance and a good dose of hope.

To exacerbate these uncertainties, the training itself tends to be a harsh environment. Many musicians go into music out of love but become disheartened by the training that oftentimes beelines to what the student did wrong or imperfectly. It does not encourage the best in people when they open and are vulnerable, only to have their imperfections exposed immediately afterward. Criticism and too much negative feedback shuts down our brains rather than opening them up to learning and creativity.

There must be a better, more healthy way. Engaging the strengths is one of the most powerful ways to combat this overly negative lens. Here are some activities designed for different music populations.

Strength Activities for Individual Musicians

- Mine your strengths to accomplish practice goals. Our top strengths are the fertilized soil and environment to leverage and grow all our strengths. Make sure you are actively naming the strengths you will need and have a system of rewards on a calendar when you achieve them. Stickers and stars still go a long way, even for adults.
- Use strengths as a writing prompt in songwriting. For instance, write a song about bravery or use honesty as a filter to run over all your lyrics to make sure they are getting to the truth of what you want to say. Or write a song talking directly to one of your strengths (dear Prudence, anyone?). Another thing you can do is take your own strength order chart and reverse it – write from the perspective of someone whose signature strengths were your lowest (numbers 20 through 24). If you are high on forgiveness, how would your lyric change if it was one of your lowest?
- Identify strengths you feel you need to get you through an upcoming challenge (Bravery? Perseverance?) and turn to those strengths in the First Set. Adopt some of the strength-specific activities in the "Lights! Camera! Action!" sections of those strengths.

- As part of your character work in an opera or musical, invent a strength profile for the person you are playing. Take some time to connect the strengths with the behaviors and roles this character portrays. Is this person similar to or different from you?
- Write your strengths on your sheet music. There is nothing like a little visual boost to remind you of who you are under pressurized situations.
- Create a playlist that makes you feel in flow with your strengths. Listen to this through the day or whenever you need a boost. Create different playlists to soothe, support, or elevate you depending on what you need and how you feel.
- Perhaps pose the question, "How am I bringing myself to this music through the lens of my strengths?"

Strength Activities for Choirs and Orchestras

- Use the strength mat for this or print out individual strengths and place them on the floor in your space. Have each member of the choir or orchestra stand on their top strength and for one minute, engage in a conversation with anyone else who shares that with them. Discuss the ways that strength shows up for you, is important to you, and how it energizes you. Then everyone names their second strength. Do this for all five signature strengths, time permitting. This is a great icebreaker or way for people to get to know one another from a core-value place. Bonus – have someone keep a tally as a scattergram or other chart to see who has the same top strengths. You may find your ensemble exhibits common or very complementary strengths. Also, within sections, there may be similarities.
- You may aim to bring more individuality into a group dynamic. This may be more applicable to vocal jazz, musical theater, or dance ensembles. By learning the top strengths of those in the section or group, you can encourage each person to accent the whole with their own flair or flavor. Take time to gauge and speak about if there is a felt difference in the outcome or if the participants feel more engaged.
- The music itself or the tone of an entire program can present strength opportunities. If a piece seems to exhibit a guiding strength, like beauty or perseverance, or kindness, send members home with an assignment to journal about the strengths identified in the music. One good basic writing prompt is, "How has *this strength (name it)* shown itself in your life? In your musicianship? In your relationships? In your thoughts and feelings?"
- Create a 30-day strengths challenge and post to your socials and have people join you! Pick one strength for everyone to boost – be it beauty, zest, etc. Ask specific questions about what people find beautiful in their

life or day. Or identify people, places, and events that heighten their zest. You can bring this into a beautiful or exciting piece of music and have the group mindfully reflect on their answers before performing. Getting others involved creates activation and is called the "contagion effect."

- Cultivate Love of Learning. Bring awareness to a historical or programmatic piece by having members engage in research. Gamify it by having a contest to find all the facts they can about the composition or history on which it was based. Have a lightning round for the first five minutes of rehearsal where members say all the things they learned. Give bonus points to anyone who brings in a lesser-known fact that no one else found. Then start rehearsal. This is a wonderful primer: it energizes, connects, and engages everyone even before you begin playing.

- If a piece conjures strength, send members home thinking about times they were strong and have them journal about it. Encourage them to boil this down to one word that they will then write on their sheet music during rehearsal. Gauge if you feel a difference in the passion of that performance. Weave this in by taking the time to ask for a few one-word shares from this experience. Remember to check in with the group and ask them to gauge if they feel a difference in the energy of that performance, how would they describe the difference?

- Use humor as an icebreaking tool. It could be as simple as putting a meme up on a projector to purposefully break the tension. Find several of these that might be applicable to different situations you know you may encounter with your group. There are tons of funny music/ musician jokes that you can have at the ready. This also works in an executive situation.

- Introduce yourself to your group through the lens of your own strengths (to see an example of this, turn back to "Backstage: About the Authors"). Then, after having members take the VIA assessment, have them turn to those next to them and do the same. By the way, sometimes it is not feasible for everyone to take the VIA survey, or people forget to do it. Have a list of the 24 character strengths handy so people can still guess at their top five, allowing everyone to participate. You can take a moment to have members scatter across the room and introduce themselves to people they don't know well. Take note of the sound that emerges after the group has bonded on a deeper level.

Strength Activities for the Music Industry

- Use strengths to delegate goals. First, have everyone on the team take the VIA survey. After allowing the group to share their top five signature strengths, use this Periodic Table of Strengths to create name tags with everyone's signature strength abbreviations (Figure 4.1).

THE PERIODIC TABLE OF STRENGTHS

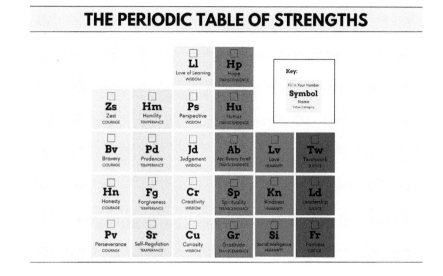

FIGURE 4.1 Periodic Table of Strengths

For instance, someone's name tag may read:

> Hn, Cu, Gr, Fr, Hp (Honesty, Curiosity, Gratitude, Fairness, Hope). This will serve to remind each member of their own and others' strengths. You can also use this chart to catalog your own strengths by filling in each rank order number in the box provided. Next, discuss and plan the ways each person's strengths can contribute towards achieving the main goal, and be as specific (name behaviors and how) as possible. Let employees choose their roles on the tasks they feel most drawn to, excited about, and naturally aligned with. Compare the outcome of this process done through a strength-lens to other tasks that were randomly assigned to employees. You may discover, as the research validates, this is a great formula for productivity and high engagement that falls along the natural lines of your employees' abilities and character strengths. We could call this "strengths-based teaming."

- When working with artists on their branding, it's important to start from a vision board or similar tool. This can sometimes be difficult for them to articulate into words. Use the artist's strengths as an entry point into deeper thought on their desired image.
- There are many tasks artists need to do on a daily basis to participate in their own business. As a manager or creative team, go through your artist's strengths and really understand them. Create detailed and very specific benchmarks using their top strengths as to how they will achieve their planned goals. Create a calendar, a system of check-ins, a badge

system ... whatever it is that will motivate your client. Learn and refer to their signature strengths regularly to help them leverage these energizers to power through.

- Build in time and space for mindfulness practices in business operations. This is a real statement about your values of focus, calm, and presence to offset the overstimulation or chaos of most businesses today. Research supports that not only do people feel better, and have better relationships, but they also are more productive and creative. Allowing a pause to take even three deep breaths will help reset a room. Infuse a one-minute break with a personal inventory of everyone's goals and how they are bringing their best self to those.

- Have a strength challenge around one of your lower strengths and have each person choose a strength they would like to build and optimize. Use this book for ideas! Journal/plan at least three ways they will work on building that strength. Check-in with each person after a designated amount of time.

- Have interviewees take the VIA survey. It is a unique way to ask people to talk about their strengths in more specific useful terms. Not only will you get to know candidates on a different level, but you also get to ask questions that will yield more personal answers. For instance, you can ask how their various strengths show up in their resume and more importantly, how they energize and impact their work life.

- If you are embarking on a large-scale project like putting out an album or preparing for a merger, identify the cluster of strengths that will set you up for success for that endeavor. Create a banner with this for the whole company to see daily. Send out an email with an inspirational quote or lyric about this strength. Add spontaneous photos that capture the best of people at work and play. It's amazing what can happen when people feel seen, successful, and unified collaborating around a concept.

- Identify a strength that is at the heart of a campaign or task – for example, bravery. Create a playlist of songs that inspire that energy and play it in the background while people are working. You can rotate this activity so that each person can highlight their core strength(s) and share their inspirational strengths playlist (and favorite music).

- As an artist manager, really look at your artists' strengths and see if what they are doing feels in alignment with who they are. Ask them if they feel energized, engaged, or stimulated. Are they unsuccessful because they are doing something in conflict with one of their signature strengths? It may be a business deal that does not align with their values, a branding decision that they feel is inauthentic, or an achievement they don't want that people believe they do want. Talk to your artist through the lens of their strengths. Is there another way they could use their stronger strengths and achieve the goal differently? Sometimes those conversations are deep and hard to get started. Giving your client a new language to be able to identify and talk about their strengths unearths new solutions, deepens the relationship, and makes you a more effective manager.

- Recall or find a conflict that you've had/have with someone in the business. See if you can think of another way you might look at that conflict using your knowledge of character strengths and intuiting some of theirs. If it is appropriate, you can both take the VIA survey and discuss your different profiles to get a more appreciative lens of your differences. Could you see the benefit in the other person's stance if you looked through their lens of values? Could there be a better understanding if this situation were to replay using strength-based approaches?

Strength Activities for the Classroom

- The classroom is a learning life lab where cognitive and experiential learning blends and takes creative forms. Instead of simply taking attendance, have each student check in by naming a strength they used the day before. Keep the strengths visible on the board or on a poster in the room and encourage all your students to keep them on their person. Encourage students to use strengths outside of the 24-strength classification by naming their own and adding to a grand list.
- Have students design customized lessons filtering through their personal strengths profile. For instance, if a student has high appreciation of beauty or excellence, how could they apply that to feel most engaged? Perhaps they want to spend time taking beautiful notes on pretty paper or with fun pens? Would they learn best by a window? If a student is high in humor, would they like to turn one unit into a funny skit or take notes in cartoon form? Let the imaginations of your students run wild and remove the boundaries. One class dedicated to this type of inquiry may leave you with a lifetime of new revelations.
- Use the strength mat (or make one of your own) in creative ways in different lessons. Employ the virtue categories to discuss and organize different elements of music history, theory, and current events. For instance, the justice category could be a jump-off point for a legal musical discussion. The humanity category could fuel a social justice music discussion. Get creative! Also, having people stand on powerful words like *fairness* and *love* gets students more engaged.
- Organize a classroom activity by students' strengths. Let them choose the tasks they feel a natural pull towards and encourage their momentum with strength spotting along the way.
- Humor is a great classroom tool! Have students create funny memes for subjects they find most difficult, like music theory. Lighten the fear factor and see resilience grow!

Strength Activities for Bands and Chamber Groups

- Get to know your smaller ensemble (rock band, chamber group, barbershop quartet) through your strengths. These closely knit musical working relationships are preciously important. It's a hugely important distinction that makes a difference when you can view each other's personalities as unique strength combinations rather than simply differences. Take the VIA survey together. Spend one whole rehearsal talking with each other about your findings. You will be amazed at the benefits you will get both during smooth sailing and rough seas. Appreciating each person's gifts and aptitudes will create a more well-rounded and intimate end product.

- Redefine conflicts amongst your members by looking at each person's strengths. For example, it annoyed you that one member of the group took forever to finish the album art. Upon further inspection, you see they have an appreciation of beauty and fairness in their top strengths. They wanted it to not only be beautiful and perfect but wanted to make sure each member of the group thought so as well. Do you see how this just got less annoying and sparked a more empathetic response?

- Remember, these strengths are here to pull you through difficulties. And working in small groups creates a team of strengths. Lean on someone else who has the strength you need at the moment. Stage fright? The trombonist has bravery as his top strength – use him for a pre-performance pep talk. Burn out? Your lead singer is a queen of perseverance. Take some coaching tips from her daily behaviors.

- In a band, make sure that you intimately know each other's top and bottom strengths. Commit to being a team and understand that your puzzle pieces move together in an interlocking machine. Anticipate where someone may falter because of a lower strength and compensate with someone else's higher strength. Reexamine old conflicts to see if they were caused by someone inadvertently violating another's core value or signature strength. It's easier to understand people's actions when you understand their motivations and what is dearly important to them.

- For fun, personify each person's musical part through their top strength. What would it sound like to play the bass line with humor? Or play the viola part with zest? Have fun with this!

- Write a love letter to your differences. Understanding and appreciating differences between people is not only important in group work, but as a global citizen. This can be especially useful in duos and in small teams where conflicts might be felt most strongly. Find things you admire and see as advantages in your differences and get to writing! Read your letter aloud to yourself or share it with others.

Strength Activities for Therapeutic Uses

- Design music therapy interventions with your clients' strengths in mind. This has the potential to encourage, empower, and unlock possibilities.
- Use music to bring forth a song that feels like or directly mentions a person's top strength. Use this song for tasks that are extra challenging.
- Make a playlist of strength-themed songs. Have patients add to it and use it for activities requiring that strength to show up.
- Be present in your own strengths. Introduce yourself to your patients through your strengths and allow them to introduce themselves to you in the same way. This is a really great way to share things that make you fulfilled and feel in flow.
- Print out all the strengths and have your client's signature strengths visible in the treatment room during their session. This is a great way to personalize and connect in the background of other tasks. Also, it's a great reminder to you as a therapist of methods that may reach them more easily or naturally.

Strength Challenge! Going Deeper with Individual Strengths

To amp up the excitement and creativity around using your strengths, alone or with a group, here are some power practices for all 24 strengths. Use these with timed or themed challenges. Focus on these at your own pace – it could be a daily, weekly, or monthly challenge that you choose do to with your team, band, co-workers, or even yourself.

- Helpful hint: Season these best practices well with action verbs that operationalize the strengths. Think of these verbs like energizer bunnies that help you put abstract concepts into use that aren't directly observable like spirituality or humility. They will help you focus on behaviors that cultivate good habits that facilitate your journey.

Appreciation of Beauty and Excellence (Transcendence)

Look around your living space or go outside in nature and focus on something that is beautiful to you for at least ten minutes. Notice every detail and use all your senses (sight, smell, touch, auditory, kinesthetic). Nature is full of this strength at every moment. Listen to your favorite piece of music or song and really hear the extraordinary notes, tones, and genius that went into creating this. Notice the positive emotions that arise: awe, joy, inspiration, love, serenity, and hope.

Bravery (Courage)

When needing to activate your bravery, make a projection into the future and think of all the benefits and outcomes of your brave actions. How will you feel when you do them? Put your fear "in your backpack" – throw it over the wall – and act "as if." As you repeat this bravery, the "as if" falls away, and you are in sync – doing and feeling brave. The "as if" is just as real in your brain and warms you up to be brave. Think of a personal fear – but not your biggest one to begin. Name three baby steps towards getting through it. Just do one thing toward it.

Creativity (Wisdom)

When doing a task or facing a problem, think of multiple and divergent ways to address it. Explode the possibilities and don't rule anything out in the beginning rounds. Think of one of your problems – a smaller one, please! Imagine three "out-of-the-box" responses. Don't think of your plan of action just yet. Take time to just explore them with a 360-degree lens.

Curiosity (Wisdom)

Ask every question you can think of that will get to what is unusual about this experience or person. Follow your intuition wherever it leads you. Think in nonlinear ways. Pay attention to every little detail. How many novel and unique aspects can you discover? As part of your regular day, take a different route to where you are going for the whole day. Select one person each day to talk to and be very curious. Ask them at least three questions about themselves, their lives, and what they like.

Fairness (Justice)

Whenever you act or speak, pause and think about how you want to be treated, spoken to, and ultimately show up in the world. Words create worlds and reality. Go out of your way to be inclusive and make room for everyone to speak. Draw out the people who are ignored or are newcomers, either in conversation or by sitting next to them.

Forgiveness (Temperance)

This is all about knowing that we are all human – which means mistakes are part of the human experience of life! "Permission to be human" is one of the best gifts you can give yourself and others. Put things in perspective, and don't sweat the small stuff – let it go. Forgiveness is for you – not just the other person. It makes life easier and lighter.

Gratitude (Transcendence)

This is the best go-to strength that always fills your cup and gets you out of your down times. Think about all the people and experiences in your life that make you who you are and brought you to this time and place. We never do it alone. Gratitude is not a word, a phrase, or a Hallmark card – it is a way to see your life that gives perspective and always connects you to others with appreciation. At least three times a week, tell someone how and what you are grateful for about them. Leave a surprise note or card for someone, or express appreciation to someone who deserves it but you don't typically tell.

Honesty (Courage)

This one can be a push especially if you don't like to hurt people's feelings or try to avoid conflict. Honesty with kindness or at least neutrality comes with maturity. Rather than having to redo, or come back to clean up a partial truth, think about the pros and cons of your choices. Which will leave you feeling freer and support your integrity? With avoidance and hedging, you leave a lot of cleanups and create anxiety for yourself. Honesty feels so clean and crisp – like a fresh fall breeze!

Hope (Transcendence)

This is the only character strength that grows in dark places in your life and is elicited in difficult times. Hope is your projection into the future and it "must have legs" to stand on and move. In other words, it's not wishful thinking or fantasy – you know what you want and plan for ways you will get there. Hope has "way power" and "how power." Find or watch one of your favorite movies that embodies hope.

Humility (Temperance)

This strength is a comforting superpower that solidly grounds you and helps you focus on others for their contributions. It allows you to avoid the tumult of aggressiveness and competition. You know who you are and are quietly confident in your abilities. One easy way to get here is to be quiet and spend more time listening and asking about others. Adopt a humanistic lens whereby we are all more alike than different. Decide that other people matter, as much as you and that we are all human after all.

Humor (Transcendence)

Having a sense of humor is a buffer during hard times that allows you to release tension and have a lighter perspective – especially when you're taking life and yourself too seriously. Binge-watch your favorite funny TV series. When something upsets you, re-write the script of who said what – but as a satire or comedy.

Judgment (Wisdom)

Take a 360-degree approach and collect as much information as you can about something you're exploring or dealing with. Rule nothing out and do not evaluate anything. Be open-minded with an attitude of "teach me and tell me more." When you hear an opposite point of view, whether personal or political, suspend your own beliefs and walk in the other's shoes. Be able to state their viewpoint.

Kindness (Humanity)

Kindness feels like a cuddly blanket or a warm cup of your favorite tea. Have the image of the words coming from your heart and going directly to the heart of the other person or people. Look at others through soft eyes. Repeat often to yourself, "Other people matter." Even in difficult times, imagine you are speaking with someone who deserves your respect as another humankind on the planet.

Leadership (Justice)

You are naturally drawn to organize people towards a common goal, and you work to bring out the best in the people. Listen to people and what they like and are interested in. Let them choose their own roles and whom they like to work with – they will be more engaged and enjoy the process. Allow space and time for people to be creatively part of the planning. This way, everyone will have "skin in the game."

Love (Humanity)

Love is its own golden energy that lights, lifts, supports, protects, and expands our relationships. Think of people in your life who love you, have loved you – whether here or gone. Choose "love coaches and mentors" from life, books, or movies that are role models of how to speak, act, and treat people. Think of the many different kinds of love – filial, agape, world, pets, animals, and romantic love. Your heart is a big place, it has lots of room to love in different ways. Love = expression = action. Share love every day in some way. Leave a love note, text, or call someone in your life today to let them know they are loved.

Love of Learning (Wisdom)

The world, people, and experiences are all catalysts for learning. Go on a deep dive google search or even better, go to a bookstore or library and let yourself wander just based on your interests and mood at the time. Just for fun, pick something new that you heard about today and spend 20 minutes learning about it. Let your curiosity take you by the hand into an unending land of discovery and learning, for its own sake.

Perseverance (Temperance)

Giving up is not an option – you persist to reach your goals. You see failures and setbacks as part of the normal process and opportunities to learn and provide useful information – then you reset. Reframing is a skill you have honed, and everything becomes a learning opportunity to aid your growth. This is an amazing and gentler way to live. Bring it on!

Perspective (Wisdom)

Go to the top of the mountain, or the highest row of the theater, and look down at the ground or stage from this wider point of view. Look at "the whole" – its shape, outline, and contour. What is the overarching theme or message? Summarize the main points. This is your ability to put the 500-piece puzzle together and finally see, "Oh, it's a picture of the Grand Canyon!" This is the wisdom of being able to see all the details and put them together to see the big picture.

Prudence (Temperance)

This is about being cautious and thinking about things before acting. Here's a good example – when cooking, only put in a little salt at first. You can always put in more, but if it's too salty, you're stuck with it. This also allows you to practice containment and being non-reactive – sometimes, "less is more."

Self-Regulation (Temperance)

This strength has so many applications – in relationships, professions, learning, exercise, and healthy life habits like eating, exercise and sleeping. In many ways, discipline is required to succeed in life. It is linked to people who succeed in every field. Practice this in baby steps, as it's like a muscle that you have to exercise and build up. Don't give up! Instead, add small chunks of time and practice every few days, or if it is something hard, each week. This is about conditioning and creating good habits.

Social Intelligence (Humanity)

This strength allows you to "read the room and the people in it." When entering any situation use your mindfulness skills to watch, look, and listen to feel the energy around you before you jump in. Use your senses and intuit.

Spirituality (Transcendence)

Let your creativity be your guide here as spirituality can take so many different forms – beyond your religious or formal belief system. Nature, beauty, and awe are ever available and replenishable to expand, transcend, and elevate you. A fresh rain, the sun on your face, a budding new flower, the smile of a baby laughing, or someone sharing something moving. All these small experiences give your life meaning and color and connect us together. We are all more alike than we think – we are not alone. This awareness is a golden gem lighting your life. Look for the good and the golden in every day. See the "sacredness" in each moment and in every relationship.

Teamwork (Justice)

Thinking about teamwork through the lens of one-for-all, and all-for-one creates the right framework. Also, keeping the question, "How can I help?" front and center reminds us to work to bring out the best, not just in ourselves, but in others. The joy and high of a synergistic team working towards a goal is a wonderful feeling of connection that lights everyone up. "The whole is greater than the sum of its parts" is the theme of great teamwork in action.

Zest (Courage)

Being energetic and full of life is so desirable and such an obvious strength, it bubbles over into an enthusiastic attitude toward all of life. One way to expand this is to take up something physical – running, dancing, kickboxing – anything you are drawn to. Watch children or puppies at play and the spontaneity of their random actions. Zest can take many different forms – you can express it through your speaking, how you dress, or even wearing a lot of colors and "happy accessories." Bring some zest to your day!

ENCORE

Encore

1. *noun*: an extra song or set after at the end of a show; one last moment the band has already left the stage   to be with the audience.

An encore has the power to take the joy and energy higher, one last time, while the performer and audience nostalgically acknowledge the bittersweetness of the show ending and therefore the value of that one last moment together. As we bring this character strength odyssey to a close, we emphasize the power of proudly and generously always leaning on your strengths through both wonderful and trying times. Let this be a changing moment in the way you look at yourself and others around you.

So how do you keep your own music playing? Go back through this book often. Reflect on situations you have had in the past, things you are going through now, and your future goals. See all the ways strengths are always operating within and around you. Harness this power! Knowing yourself in this new way will surely open doors, clarify decisions, and help you truly savor your own special strength recipe.

Here are some closing prompts to summarize what you have learned about yourself and your musical entity. These are inspired by the work of Dr. Margarita Tarragona's end-of-year summary (Tarragona, 2020).

- What was your favorite part of this book?
- What did you enjoy most about learning about your strengths?
- What is something important you learned about a collaborator?
- What was your favorite ah-ha moment?
- What is a strength you want to cultivate more of?
- What would you like to manifest using your strengths?
- What would you like to leave behind that no longer serves you?
- What strength will bring you closer to your dreams this year?

Each time we teach and show people their character strengths, it's like watching a new birth. Reframing yourself into these neutral positive virtue categories and strengths at first takes some getting used to. However, as time goes on, it's like watching an inspiring song get written, produced, and proudly played.

In the name of prioritizing your positivity through your unique strengths profile, we hope you are inspired to keep playing and sharing your song.

Last Set Practice Room

Take an activity you currently do in your professional/musical life. Redesign some aspect of it specifically around using or integrating your character strengths. Next, identify one of your lower strengths (17–24) that could help you in your current situation today. Use one of the activities defined in the FIRST SET to cultivate and elevate this under-used strength. How did the introduction of your lower strengths impact the experience? How did you leverage one of your stronger strengths to help the lower strength engage?

_____

_____

_____

_____

_____

_____

_____

_____

_____

_____

AFTERPARTY

Afterparty

1.   *noun*: a party after a big event or concert; a joyous way to continue celebrating.

Still want more? Here is a categorized list of some helpful resources to continue your positivity and character work.

Retreats and Educational Institutes

- The VIA Institute: https://viacharacter.org
- Positive Psychology by Barbara L. Fredrickson on Coursera
- Whole Being Institute: https://wholebeinginstitute.com
- Greater Good Science Center at UC Berkeley: https://greatergood. berkeley.edu
- The International Storytelling Festival: https://storytellingcenter.net
- World Happiness Summit: https://worldhappinesssummit.com
- Esalen Human Potential Retreat: https://esalen.org
- Omega Center for Sustainable Living: https://eomega.org
- Kripalu Center for Yoga and Health: https://kripalu.org
- Happiness Studies Academy: https://talbenshahar.com
- Neurodharma Online Program: https://rickhanson.net
- Movement and Yoga with your character strengths: https://letyouryoga-dance.com
- Many other free positive psychology courses online taught by experts in the field on Coursera, Udemy and EdX

Inspirational and Informative TED Talks

- Developing a Growth Mindset with Carol Dweck
- The New Era of Positive Psychology with Martin Seligma
- How to Make Stress Your Friend with Kelly McGonigal
- Flow, the Secret to Happiness with Mihaly Csikszentmihaly
- The Happy Secret to Better Work with Shawn Acho
- The Secret to Mastering Life's Biggest Transitions with Bruce Feiler
- Hardwiring Happiness with Rick Hanson
- The Key to Success, Grit with Angela Lee Duckworth
- The Surprising Science of Happiness with Dan Gilber
- Warning: Being Positive is not for the Faint-Hearted! With Lea Waters
- The Power of Vulnerability with Brene Brown

- How Positive Emotions Work and Why by Barbara L. Fredrickson
- The Surprising Power of Original Thinkers with Adam Grant
- Success, Failure, and the Drive to Keep Creating with Elizabeth Gilbert
- Remaking Love with Barbara Fredrickson
- A Universal Language that Describes What's Best in Us with Ryan Niemiec

MERCH BOOTH

MERCH BOOTH

1. *noun*: a place where vendors can set up merchandise relating to the show; a place where fans can buy items that will remind them of the experience they just had.

Helpful Tools to Continue Your Strengths Practices

- Segura Strength Cluster Mat: https://www.gisellemarzosegura.com/strength-clusters
- Character Strength Cards Moo.com on www.positivepsychologyfor-musicprofessionals.com
- Custom Strength Bracelets from DashandDotShop https://etsy.com/shop/DashandDotShop

References

Asplund, J., Elliot, G., & Flade, P. (2023, January 23). *Employees who use their strengths outperform those who don't.* Gallup.com. Retrieved February 13, 2023, from https://www.gallup.com/workplace/236561/employees-strengths-outperform-don.aspx

Budson, A. E. (2020, October 7). *Why is music good for the brain?* Harvard Health. Retrieved January 10, 2023, from https://www.health.harvard.edu/blog/why-is-music-good-for-the-brain-2020100721062

Build your confidence by identifying your strengths – weidel on winning. Weidel on Winning. (2022, July 26). Retrieved February 13, 2023, from https://weidelonwinning.com/blog/build-your-confidence-by-identifying-your-strengths/

Clarke, S. (2018, May 17). *Playing to strengths – do your people do what they love, every single day?* LinkedIn. Retrieved February 13, 2023, from https://www.linkedin.com/pulse/playing-strengths-do-your-people-what-love-every-single-clarke/

Dalla-Camina, M. (2022, November 18). *How strengths fuel your confidence.* Women Rising. Retrieved February 13, 2023, from https://womenrisingco.com/articles/how-strengths-fuel-your-confidence/

Focus on your strengths, focus on success. It's Your Yale (n.d.). Retrieved February 13, 2023, from https://your.yale.edu/work-yale/learn-and-grow/focus-your-strengths-focus-success

Haverson, V. I. (2018, June 6). *How to find happiness through your strengths.* Ellevate. Retrieved February 13, 2023, from https://www.ellevatenetwork.com/articles/9122-how-to-find-happiness-through-your-strengths

McKay, B., & McKay, K. (2021, September 25). *Building your resiliency: Part V – recognizing and utilizing your signature strengths.* The Art of Manliness. Retrieved February 13, 2023, from https://www.artofmanliness.com/character/behavior/building-your-resiliency-part-v-recognizing-and-utilizing-your-signature-strengths/

Roberts, L. M., Barker, B., Heaphy, E. D., Quinn, R. E., Dutton, J. E., & Spreitzer, G. (2022, January 4). *How to play to your strengths.* Harvard Business Review. Retrieved February 13, 2023, from https://hbr.org/2005/01/how-to-play-to-your-strengths

Rothstein, L., & Stromme, D. (2018, December 11). *Let your strengths fuel your energy.* Positive Psychology | UMN Extension. Retrieved February 13, 2023, from https://extension.umn.edu/two-you-video-series/let-your-strengths-fuel-your-energy

Tarragona, M. (2020, September 29). *A New Year's ritual.* Wholebeing Institute. Retrieved February 13, 2023, from https://wholebeinginstitute.com/a-new-years-ritual/

Wood, A. M., Geraghty, A. W. A., Brdar, I., Cohen, S., Cook, T. D., Duckworth, A. L., Govindji, R., Harter, J. K., Lakey, B., & Linley, P. A. (2010, September 15). *Using personal and psychological strengths leads to increases in well-being over time: A longitudinal study and the development of the Strengths Use Questionnaire.* Personality and Individual Differences. Retrieved February 13, 2023, from https://www.sciencedirect.com/science/article/abs/pii/S0191886910003946

Zhang, S.-e, Yang, L.-bin, Zhao, C.-xi, Shi, Y., Wang, H.-ni, Zhao, X., Wang, X.-he, Sun, T., & Cao, D.-pin. (2021, September 13). *Contribution of character strengths to psychology stress, sleep quality, and subjective health status in a sample of Chinese nurses.* Frontiers. Retrieved February 13, 2023, from https://www.frontiersin.org/articles/10.3389/fpsyg.2021.631459/full

INDEX